Essentials of Applied Mathematics for Engineers and Scientists

SECOND EDITION

Synthesis Lectures on Mathematics and Statistics

Editor
Steven G. Krantz, *Washington University, St. Louis*

Essentials of Applied Mathematics for Engineers and Scientists, SECOND EDITION
Robert G. Watts
2012

Chaotic Maps: Dynamics, Fractals, and Rapid Fluctuations
Goong Chen and Yu Huang
2011

Matrices in Engineering Problems
Marvin J. Tobias
2011

The Integral: A Crux for Analysis
Steven G. Krantz
2011

Statistics is Easy! Second Edition
Dennis Shasha and Manda Wilson
2010

Lectures on Financial Mathematics: Discrete Asset Pricing
Greg Anderson and Alec N. Kercheval
2010

Jordan Canonical Form: Theory and Practice
Steven H. Weintraub
2009

The Geometry of Walker Manifolds
Miguel Brozos-Vázquez, Eduardo García-Río, Peter Gilkey, Stana Nikcevic, and Rámon
Vázquez-Lorenzo
2009

An Introduction to Multivariable Mathematics
Leon Simon
2008

Jordan Canonical Form: Application to Differential Equations
Steven H. Weintraub
2008

Statistics is Easy!
Dennis Shasha and Manda Wilson
2008

A Gyrovector Space Approach to Hyperbolic Geometry
Abraham Albert Ungar
2008

Essentials of Applied Mathematics for Engineers and Scientists, SECOND EDITION

Robert G. Watts

ISBN: 978-3-031-01276-1   paperback
ISBN: 978-3-031-02404-7   ebook

DOI 10.1007/978-3-031-02404-7

A Publication in the Springer series
SYNTHESIS LECTURES ON MATHEMATICS AND STATISTICS

Lecture #12
Series Editor: Steven G. Krantz, *Washington University, St. Louis*
Series ISSN
Synthesis Lectures on Mathematics and Statistics
Print 1938-1743   Electronic 1938-1751

Essentials of Applied Mathematics for Engineers and Scientists

SECOND EDITION

Robert G. Watts

Tulane University and United States Naval Academy

SYNTHESIS LECTURES ON MATHEMATICS AND STATISTICS #12

ABSTRACT

The second edition of this popular book on practical mathematics for engineers includes new and expanded chapters on perturbation methods and theory. This is a book about linear partial differential equations that are common in engineering and the physical sciences. It will be useful to graduate students and advanced undergraduates in all engineering ?elds as well as students of physics, chemistry, geophysics and other physical sciences and professional engineers who wish to learn about how advanced mathematics can be used in their professions. The reader will learn about applications to heat transfer, fluid flow and mechanical vibrations. The book is written in such a way that solution methods and application to physical problems are emphasized. There are many examples presented in detail and fully explained in their relation to the real world. References to suggested further reading are included. The topics that are covered include classical separation of variables and orthogonal functions, Laplace transforms, complex variables and Sturm-Liouville transforms. This second edition includes two new and revised chapters on perturbation methods, and singular perturbation theory of differential equations.

KEYWORDS

Engineering mathematics, separation of variables, orthogonal functions, Laplace transforms, complex variables and Sturm-Liouville transforms, differential equations, perturbation methods, perturbation theory

Contents

CHAPTER 1

Partial Differential Equations in Engineering

1.1 INTRODUCTORY COMMENTS

This book covers the material presented in a course in applied mathematics that is required for first-year graduate students in the departments of Chemical and Mechanical Engineering at Tulane University. A great deal of material is presented, covering boundary value problems, complex variables, and Fourier transforms. Therefore the depth of coverage is not as extensive as in many books. Our intent in the course is to introduce students to methods of solving linear partial differential equations. Subsequent courses such as conduction, solid mechanics, and fracture mechanics then provide necessary depth.

The reader will note some similarity to the three books, *Fourier Series and Boundary Value Problems, Complex Variables and Applications*, and *Operational Mathematics*, originally by R. V. Churchill. The first of these has been recently updated by James Ward Brown. The current author greatly admires these works, and studied them during his own tenure as a graduate student. The present book is more concise and leaves out some of the proofs in an attempt to present more material in a way that is still useful and is acceptable for engineering students.

First we review a few concepts about differential equations in general.

1.2 FUNDAMENTAL CONCEPTS

An *ordinary differential equation* expresses a dependent variable, say u, as a function of one independent variable, say x, and its derivatives. The *order* of the differential equation is given by the order of the highest derivative of the dependent variable. A boundary value problem consists of a differential equation that is defined for a given range of the independent variable (*domain*) along with conditions on the boundary of the domain. In order for the boundary value problem to have a unique solution the number of boundary conditions must equal the order of the differential equation. If the differential equation and the boundary conditions contain only terms of first degree in u and its derivatives the problem is *linear*. Otherwise it is *nonlinear*.

A *partial differential equation* expresses a dependent variable, say u, as a function of more than one independent variable, say x, y, and z. Partial derivatives are normally written as $\partial u / \partial x$. This is the first-order derivative of the dependent variable u with respect to the independent variable x. Sometimes we will use the notation u_x or when the derivative is an ordinary derivative we use u'. Higher order derivatives are written as $\partial^2 u / \partial x^2$ or u_{xx}. The order of the differential equation now depends on the orders of the derivatives of the dependent variables in terms of each of the independent variables. For example, it may be of order m for the x variable and of order n for the y variable. A boundary value problem consists of a partial differential equation defined on a domain in the space of the independent variables, for example the x, y, z space, along with conditions on the boundary. Once again, if the partial differential equation and the boundary conditions contain only terms of first degree in u and its derivatives the problem is linear. Otherwise it is nonlinear.

A differential equation or a boundary condition is *homogeneous* if it contains only terms involving the dependent variable.

Examples
Consider the ordinary differential equation

$$a(x)u'' + b(x)u = c(x), \quad 0 < x < A. \tag{1.1}$$

Two boundary conditions are required because the order of the equation is 2. Suppose

$$u(0) = 0 \quad \text{and} \quad u(A) = 1. \tag{1.2}$$

The problem is linear. If $c(x)$ is not zero the differential equation is nonhomogeneous. The first boundary condition is homogeneous, but the second boundary condition is nonhomogeneous.

Next consider the ordinary differential equation

$$a(u)u'' + b(x)u = c \quad 0 < x < A \tag{1.3}$$

Again two boundary conditions are required. Regardless of the forms of the boundary conditions, the problem is nonlinear because the first term in the differential equations is not of first degree in u and u'' since the leading coefficient is a function of u. It is homogeneous only if $c = 0$.

Now consider the following three partial differential equations:

$$u_x + u_{xx} + u_{xy} = 1 \tag{1.4}$$

$$u_{xx} + u_{yy} + u_{zz} = 0 \tag{1.5}$$

$$u u_x + u_{yy} = 1 \tag{1.6}$$

The first equation is linear and nonhomogeneous. The third term is a *mixed partial derivative*. Since it is of second order in x two boundary conditions are necessary on x. It is first order in y, so that only one boundary condition is required on y. The second equation is linear and homogeneous and is of second order in all three variables. The third equation is nonlinear because the first term is not of first degree in u and u_x. It is of order 1 in x and order 2 in y.

In this book we consider only linear equations. We will now derive the partial differential equations that describe some of the physical phenomena that are common in engineering science.

Problems

Tell whether the following are linear or nonlinear and tell the order in each of the independent variables:

$$u'' + xu' + u^2 = 0$$

$$\tan(y)u_y + u_{yy} = 0$$

$$\tan(u)u_y + 3u = 0$$

$$u_{yyy} + u_{yx} + u = 0$$

1.3   THE HEAT CONDUCTION (OR DIFFUSION) EQUATION

1.3.1   Rectangular Cartesian Coordinates

The conduction of heat is only one example of the diffusion equation. There are many other important problems involving the diffusion of one substance in another. One example is the diffusion of one gas into another if both gases are motionless on the macroscopic level (no convection). The diffusion of heat in a motionless material is governed by Fourier's law which states that heat is conducted per unit area in the negative direction of the temperature gradient in the (vector) direction **n** in the amount $\partial u/\partial n$, that is

$$q^n = -k\partial u/\partial n \tag{1.7}$$

where q^n denotes the heat flux in the n direction (not the nth power). In this equation u is the local temperature and k is the thermal conductivity of the material. Alternatively u could be the partial fraction of a diffusing material in a host material and k the diffusivity of the diffusing material relative to the host material.

Consider the diffusion of heat in two dimensions in rectangular Cartesian coordinates. Fig. 1.1 shows an element of the material of dimension Δx by Δy by Δz. The material has a specific heat c and a density ρ. Heat is generated in the material at a rate q per unit volume. Performing a heat balance on the element, the time (t) rate of change of thermal energy within the element, $\rho c \Delta x \Delta y \Delta z \partial u/\partial t$ is equal to the rate of heat generated within the element

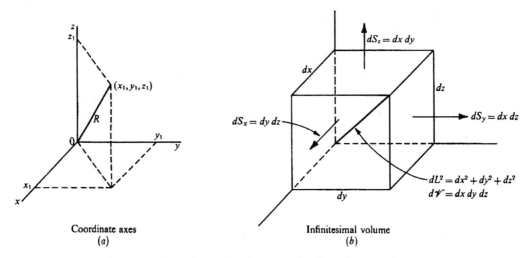

FIGURE 1.1: An element in three dimensional rectangular Cartesian coordinates

$q''' \Delta x \Delta y \Delta z$ minus the rate at which heat is conducted out of the material. The flux of heat conducted into the element at the x face is denoted by q^x while at the y face it is denoted by q^y. At $x + \Delta x$ the heat flux (i.e., per unit area) leaving the element in the x direction is $q^x + \Delta q^x$ while at $y + \Delta y$ the heat flux leaving in the y direction is $q^y + \Delta q^y$. Similarly for q^z. Expanding the latter three terms in Taylor series, we find that $q^x + \Delta q^x = q^x + q_x^x \Delta x + (1/2)q_{xx}^x (\Delta x)^2$ + terms of order $(\Delta x)^3$ or higher order. Similar expressions are obtained for $q^y + \Delta q^y$ and $q^z + \Delta q^z$ Completing the heat balance

$$\rho c \, \Delta x \Delta y \Delta z \partial u / \partial t = q''' \Delta x \Delta y \Delta z + q^x \Delta y \Delta z + q^y \Delta x \Delta z$$
$$- (q^x + q_x^x \Delta x + (1/2)q_{xx}^x (\Delta x)^2 + \cdots) \Delta y \Delta z$$
$$- (q^y + q_y^y \Delta y + (1/2)q_{yy}^y (\Delta y)^2 + \cdots) \Delta x \Delta z \qquad (1.8)$$
$$- (q^z + q_z^z \Delta z + (1/2)q_{zz}^z (\Delta z)^2 + \cdots) \Delta x \Delta y$$

The terms $q^x \Delta y \Delta z$, $q^y \Delta x \Delta z$, and $q^z \Delta x \Delta y$ cancel. Taking the limit as Δx, Δy, and Δz approach zero, noting that the terms multiplied by $(\Delta x)^2$, $(\Delta y)^2$, and $(\Delta z)^2$ may be neglected, dividing through by $\Delta x \Delta y \Delta z$ and noting that according to Fourier's law $q^x = -k \partial u / \partial x$, $q^y = -k \partial u / \partial y$, and $q^z = -k(\partial u / \partial z)$ we obtain the time-dependent heat conduction equation in three-dimensional rectangular Cartesian coordinates:

$$\rho c \, \partial u / \partial t = k(\partial^2 u / \partial x^2 + \partial^2 u / \partial y^2) + q \qquad (1.9)$$

The equation is first order in t, and second order in both x and y. If the property values ρ, c and k and the heat generation rate per unit volume q are independent of the dependent

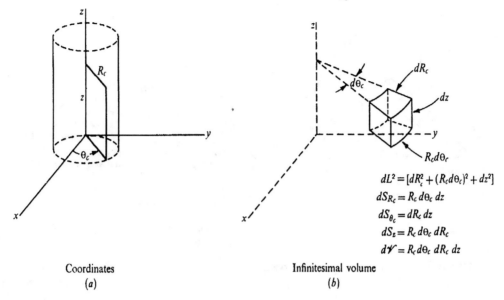

$$dL^2 = [dR_c^2 + (R_c d\Theta_c)^2 + dz^2]$$
$$dS_{R_c} = R_c \, d\Theta_c \, dz$$
$$dS_{\theta_c} = dR_c \, dz$$
$$dS_z = R_c \, d\Theta_c \, dR_c$$
$$d\mathcal{V} = R_c \, d\Theta_c \, dR_c \, dz$$

Coordinates
(a)

Infinitesimal volume
(b)

FIGURE 1.2: An element in cylindrical coordinates

variable, temperature the partial differential equation is linear. If q is zero, the equation is homogeneous. It is easy to see that if a third dimension, z, were included, the term $k\partial^2 u/\partial z^2$ must be added to the right-hand side of the above equation.

1.3.2 Cylindrical Coordinates

A small element of volume $r\Delta\Theta\Delta r\Delta z$ is shown in Fig. 1.2.

The method of developing the diffusion equation in cylindrical coordinates is much the same as for rectangular coordinates except that the heat conducted into and out of the element depends on the area as well as the heat flux as given by Fourier's law, and this area varies in the r-direction. Hence the heat conducted into the element at r is $q^r r\Delta\Theta\Delta z$, while the heat conducted out of the element at $r + \Delta r$ is $q^r r\Delta\Theta\Delta z + \partial(q^r r\Delta\Theta\Delta z)/\partial r(\Delta r)$ when terms of order $(\Delta r)^2$ are neglected as Δr approaches zero. In the z- and θ-directions the area does not change. Following the same procedure as in the discussion of rectangular coordinates, expanding the heat values on the three faces in Tayor series', and neglecting terms of order $(\Delta\Theta)^2$ and $(\Delta z)^2$ and higher,

$$\rho c r\Delta\theta\Delta r\Delta z\partial u/\partial t = -\partial(q^r r\Delta\theta\Delta z)/\partial r\Delta r - \partial(q^\theta \Delta r\Delta z)/\partial\theta\Delta\theta$$
$$- \partial(q^z r\Delta\theta\Delta r)/\partial z\Delta z + qr\Delta\theta\Delta r\Delta z \qquad (1.10)$$

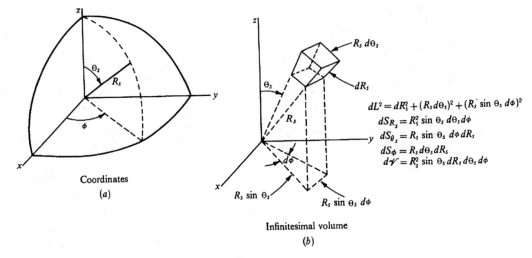

FIGURE 1.3: An element in spherical coordinates

Dividing through by the volume, we find after using Fourier's law for the heat fluxes

$$\rho c \, \partial u/\partial t = (1/r)\partial(r\partial u/\partial r)/\partial r + (1/r^2)\partial^2 u/\partial\theta^2 + \partial^2 u/\partial z^2 + q \qquad (1.11)$$

1.3.3   Spherical Coordinates

An element in a spherical coordinate system is shown in Fig. 1.3. The volume of the element is $r\sin\theta\,\Delta\Phi\,\Delta r r\,\Delta\theta = r^2\sin\theta\,\Delta r\,\Delta\theta\,\Delta\Phi$. The net heat flows out of the element in the r, θ, and Φ directions are respectfully

$$q^r r^2 \sin\theta\,\Delta\theta\,\Delta\Phi \qquad (1.12)$$

$$q^\theta r \sin\theta\,\Delta r\,\Delta\Phi \qquad (1.13)$$

$$q^\Phi r \,\Delta\theta\,\Delta r \qquad (1.14)$$

It is left as an exercise for the student to show that

$$\rho c \, \partial u/\partial t = k[(1/r^2)\partial/\partial r(r^2\partial u/\partial r) + (1/r^2\sin^2\theta)\partial^2 u/\partial\Phi^2$$
$$+ (1/r^2\sin\theta)\partial(\sin\theta\,\partial u/\partial\theta)/\partial\theta + q \qquad (1.15)$$

The Laplacian Operator

The linear operator on the right-hand side of the heat equation is often referred to as the Laplacian operator and is written as ∇^2.

1.3.4   Boundary Conditions

Four types of boundary conditions are common in conduction problems.

   a) Heat flux prescribed, in which case $k\partial u/\partial n$ is given.

   b) Heat flux is zero (perhaps just a special case of (a)), in which case $\partial u/\partial n$ is zero.

   c) Temperature u is prescribed.

   d) Convection occurs at the boundary, in which case $k\partial u/\partial n = h(U - u)$.

Here n is a length in the direction normal to the surface, U is the temperature of the fluid next to the surface that is heating or cooling the surface, and h is the coefficient of convective heat transfer. Condition (d) is sometimes called Newton's law of cooling.

1.4   THE VIBRATING STRING

Next we consider a tightly stretched string on some interval of the x-axis. The string is vibrating about its equilibrium position so that its departure from equilibrium is $y(t, x)$. The string is assumed to be perfectly flexible with mass per unit length ρ.

Fig. 1.4 shows a portion of such a string that has been displaced upward. We assume that the tension in the string is constant. However the direction of the tension vector along the string varies. The tangent of the angle $\alpha(t, x)$ that the string makes with the horizontal is given by the slope of the wire, $\partial y/\partial x$,

$$V(x)/H = \tan \alpha(t, x) = \partial y/\partial x \qquad (1.16)$$

If we assume that the angle α is small then the horizontal tension force is nearly equal to the magnitude of the tension vector itself. In this case the tangent of the slope of the wire

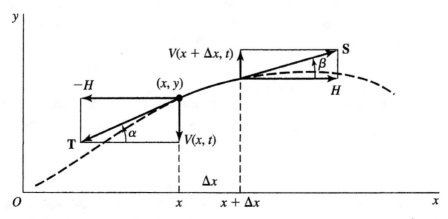

FIGURE 1.4: An element of a vibrating string

at $x + \Delta x$ is

$$V(x + \Delta x)/H = \tan\alpha(x + \Delta x) = \partial y/\partial x(x + \Delta x). \qquad (1.17)$$

The vertical force V is then given by $H\partial y/\partial x$. The *net* vertical force is the difference between the vertical forces at x and $x + \Delta x$, and must be equal to the mass times the acceleration of that portion of the string. The mass is $\rho\Delta x$ and the acceleration is $\partial^2 y/\partial t^2$. Thus

$$\rho\Delta x\partial^2/\partial t^2 = H[\partial y/\partial x(x + \Delta x) - \partial y/\partial x(x)] \qquad (1.18)$$

Expanding $\partial y/\partial x(x + \Delta x)$ in a Taylor series about $\Delta x = 0$ and neglecting terms of order $(\Delta x)^2$ and smaller, we find that

$$\rho y_{tt} = H y_{xx} \qquad (1.19)$$

which is the wave equation. Usually it is presented as

$$y_{tt} = a^2 y_{xx} \qquad (1.20)$$

where $a^2 = H/\rho$ is a wave speed term.

Had we included the weight of the string there would have been an extra term on the right-hand side of this equation, the acceleration of gravity (downward). Had we included a damping force proportional to the velocity of the string, another negative term would result:

$$\rho y_{tt} = H y_{xx} - b y_t - g \qquad (1.21)$$

1.4.1   Boundary Conditions

The partial differential equation is linear and if the gravity term is included it is nonhomogeneous. It is second order in both t and x, and requires two boundary conditions (initial conditions) on t and two boundary conditions on x. The two conditions on t are normally specifying the initial velocity and acceleration. The conditions on x are normally specifying the conditions at the ends of the string, i.e., at $x = 0$ and $x = L$.

1.5   VIBRATING MEMBRANE

The partial differential equation describing the motion of a vibrating membrane is simply an extension of the right-hand side of the equation of the vibrating string to two dimensions. Thus,

$$\rho y_{tt} + b y_t = -g + \nabla^2 y \qquad (1.22)$$

In this equation, ρ is the density per unit area and $\nabla^2 y$ is the *Laplacian operator* in either rectangular or cylindrical coordinates.

1.6   LONGITUDINAL DISPLACEMENTS OF AN ELASTIC BAR

The longitudinal displacements of an elastic bar are described by Eq. (1.20) except the in this case $a^2 = E/\rho$, where ρ is the density and E is Young's modulus.

FURTHER READING

V. Arpaci, *Conduction Heat Transfer*. Reading, MA: Addison-Wesley, 1966.

J. W. Brown and R. V. Churchill, *Fourier Series and Boundary Value Problems*. 6th edition. New York: McGraw-Hill, 2001.

P. V. O'Neil, *Advanced Engineering Mathematics*. 5th edition. Pacific Grove, CA: Brooks/Cole-Thomas Learning, 2003.

CHAPTER 2

The Fourier Method: Separation of Variables

In this chapter we will work through a few example problems in order to introduce the general idea of *separation of variables* and the concept of orthogonal functions before moving on to a more complete discussion of orthogonal function theory. We will also introduce the concepts of *nondimensionalization* and *normalization*.

The goal here is to use the three theorems stated below to walk the student through the solution of several types of problems using the concept of separation of variables and learn some early lessons on how to apply the method without getting too much into the details that will be covered later, especially in Chapter 3.

We state here without proof three fundamental theorems that will be useful in finding series solutions to partial differential equations.

Theorem 2.1. *Linear Superposition: If a group of functions u_n, $n = m$ through $n = M$ are all solutions to some linear differential equation then*

$$\sum_{n=m}^{M} c_n u_n$$

is also a solution.

Theorem 2.2. *Orthogonal Functions: Certain sets of functions Ψ_n defined on the interval (a, b) possess the property that*

$$\int_a^b \Psi_n \Psi_m dx = constant, \ n = m$$

$$\int_a^b \Psi_n \Psi_m dx = 0, \ n \neq m$$

These are called orthogonal functions. Examples are the sine and cosine functions. This idea is discussed fully in Chapter 3, particularly in connection with Sturm–Liouville equations.

Theorem 2.3. *Fourier Series: A piecewise continuous function $f(x)$ defined on (a, b) can be represented by a series of orthogonal functions $\Psi_n(x)$ on that interval as*

$$f(x) = \sum_{n=0}^{\infty} A_n \Psi_n(x)$$

where

$$A_n = \frac{\int_{x=a}^{b} f(x) \Psi_n(x) dx}{\int_{x=a}^{b} \Psi_n(x) \Psi_n(x) dx}$$

These properties will be used in the following examples to introduce the idea of solution of partial differential equations using the concept of separation of variables.

2.1   HEAT CONDUCTION

We will first examine how Theorems 1, 2, and 3 are systematically used to obtain solutions to problems in heat conduction in the forms of infinite series. We set out the methodology in detail, step-by-step, with comments on lessons learned in each case. We will see that the mathematics often serves as a guide, telling us when we make a bad assumption about solution forms.

Example 2.1. A Transient Heat Conduction Problem

Consider a flat plate occupying the space between $x = 0$ and $x = L$. The plate stretches out in the y and z directions far enough that variations in temperature in those directions may be neglected. Initially the plate is at a uniform temperature u_0. At time $t = 0^+$ the wall at $x = 0$ is raised to u_1 while the wall at $x = L$ is insulated. The boundary value problem is then

$$\rho c u_t = k u_{xx} \quad 0 < x < L \quad t > 0 \tag{2.1}$$

$$u(t, 0) = u_1$$

$$u_x(t, L) = 0 \tag{2.2}$$

$$u(0, x) = u_0$$

2.1.1   Scales and Dimensionless Variables

When it is possible it is always a good idea to write both the independent and dependent variables in such a way that they range from zero to unity. In the next few problems we shall show how this can often be done.

We first note that the problem has a fundamental length scale, so that if we define another space variable $\xi = x/L$, the partial differential equation can be written as

$$\rho c \, u_t = L^{-2} k u_{\xi\xi} \quad 0 < \xi < 1 \quad t < 0 \tag{2.3}$$

Next we note that if we define a dimensionless time-like variable as $\tau = \alpha t/L^2$, where $\alpha = k/\rho c$ is called the *thermal diffusivity*, we find

$$u_\tau = u_{\xi\xi} \tag{2.4}$$

We now proceed to *nondimensionalize* and *normalize* the dependent variable and the boundary conditions. We define a new variable

$$U = (u - u_1)/(u_0 - u_1) \tag{2.5}$$

Note that this variable is always between 0 and 1 and is dimensionless. Our boundary value problem is now devoid of constants.

$$U_\tau = U_{\xi\xi} \tag{2.6}$$

$$U(\tau, 0) = 0$$

$$U_\xi(\tau, 1) = 0 \tag{2.7}$$

$$U(0, \xi) = 1$$

All but one of the boundary conditions are homogeneous. *This will prove necessary in our analysis.*

2.1.2   Separation of Variables

Begin by assuming $U = \Gamma(\tau)\Phi(\xi)$. Insert this into the differential equation and obtain

$$\Phi(\xi)\Gamma_\tau(\tau) = \Gamma(\tau)\Phi_{\xi\xi}(\xi). \tag{2.8}$$

Next divide both sides by $U = \Phi\Gamma$,

$$\frac{\Gamma_\tau}{\Gamma} = \frac{\Phi_{\xi\xi}}{\Phi} = \pm\lambda^2 \tag{2.9}$$

The left-hand side of the above equation is a function of τ only while the right-hand side is a function only of ξ. This can only be true if both are constants since they are equal to each other. λ^2 is always positive, but we must decide whether to use the plus sign or the minus sign. We have two ordinary differential equations instead of one partial differential equation. Solution for Γ gives a constant times either $\exp(-\lambda^2\tau)$ or $\exp(+\lambda^2\tau)$. Since we know that U is always between 0 and 1, we see immediately that we must choose the minus sign. The second ordinary differential equation is

$$\Phi_{\xi\xi} = -\lambda^2 \Phi \tag{2.10}$$

and we deduce that the two homogeneous boundary conditions are

$$\Phi(0) = 0$$
$$\Phi_\xi(1) = 0 \tag{2.11}$$

Solving the differential equation we find

$$\Phi = A\cos(\lambda\xi) + B\sin(\lambda\xi) \tag{2.12}$$

where A and B are constants to be determined. The first boundary condition requires that $A = 0$.

The second boundary condition requires that either $B = 0$ or $\cos(\lambda) = 0$. Since the former cannot be true (U is not zero!) the latter must be true. ξ can take on any of an infinite number of values $\lambda_n = (2n - 1)\pi/2$, where n is an integer between negative and positive infinity. Equation (2.10) together with boundary conditions (2.11) is called a Sturm–Liouville problem. The solutions are called *eigenfunctions* and the λ_n are called *eigenvalues*. A full discussion of Sturm–Liouville theory will be presented in Chapter 3.

Hence the apparent solution to our partial differential equation is any one of the following:

$$U_n = B_n \exp[-(2n - 1)^2\pi^2\tau/4)] \sin[\pi(2n - 1)\xi/2]. \tag{2.13}$$

2.1.3 Superposition

Linear differential equations possess the important property that if each solution U_n satisfies the differential equation and the boundary conditions then the linear combination

$$\sum_{n=1}^{\infty} B_n \exp[-(2n - 1)^2\pi^2\tau/4] \sin[\pi(2n - 1)\xi/2] = \sum_{n=1}^{\infty} U_n \tag{2.14}$$

also satisfies them, as stated in Theorem 2. Can we build this into a solution that satisfies the one remaining boundary condition? The final condition (the nonhomogeneous initial condition) states that

$$1 = \sum_{n=1}^{\infty} B_n \sin(\pi(2n - 1)\xi/2) \tag{2.15}$$

This is called a Fourier sine series representation of 1. The topic of Fourier series is further discussed in Chapter 3.

2.1.4   Orthogonality

It may seem hopeless at this point when we see that we need to find an infinite number of constants B_n. What saves us is a concept called *orthogonality* (to be discussed in a more general way in Chapter 3). The functions $\sin(\pi(2n-1)\xi/2)$ form an orthogonal set on the interval $0 < \xi < 1$, which means that

$$\int_0^1 \sin(\pi(2n-1)\xi/2)\sin(\pi(2m-1)\xi/2)d\xi = 0 \text{ when } m \neq n \tag{2.16}$$

$$= 1/2 \text{ when } m = n$$

Hence if we multiply both sides of the final equation by $\sin(\pi(2m-1)\xi/2)d\xi$ and integrate over the interval, we find that all of the terms in which $m \neq n$ are zero, and we are left with one term, the general term for the nth B, B_n

$$B_n = 2\int_0^1 \sin(\pi(2n-1)\xi/2)d\xi = \frac{4}{\pi(2n-1)} \tag{2.17}$$

Thus

$$U = \sum_{n=1}^{\infty} \frac{4}{\pi(2n-1)}\exp[-\pi^2(2n-1)^2\tau/4]\sin[\pi(2n-1)\xi/2] \tag{2.18}$$

satisfies both the partial differential equation and the boundary and initial conditions, and therefore is a solution to the boundary value problem.

2.1.5   Lessons

We began by assuming a solution that was the product of two variables, each a function of only one of the independent variables. Each of the resulting ordinary differential equations was then solved. The two homogeneous boundary conditions were used to evaluate one of the constant coefficients and the separation constant λ. It was found to have an infinite number of values. These are called *eigenvalues* and the resulting functions $\sin\lambda_n\xi$ are called *eigenfunctions*. Linear superposition was then used to build a solution in the form of an infinite series. The infinite series was then required to satisfy the initial condition, the only nonhomogeneous condition. The coefficients of the series were determined using the concept of *orthogonality* stated in Theorem 3, resulting in a Fourier series. Each of these concepts will be discussed further in Chapter 3. *For now we state that many important functions are members of orthogonal sets.*

The method would not have worked had the differential equation not been homogeneous. (Try it.) It also would not have worked if more than one boundary condition had been nonhomogeneous. We will see how to get around these problems shortly.

Problems

1. Equation (2.9) could just as easily have been written as

$$\frac{\Gamma_\tau}{\Gamma} = \frac{\Phi_{\xi\xi}}{\Phi} = +\lambda^2$$

   Show two reasons why this would reduce to the trivial solution or a solution for which Γ approaches infinity as τ approaches infinity, and that therefore the minus sign must be chosen.

2. Solve the above problem with boundary conditions

$$U_\xi(\tau, 0) = 0 \quad \text{and} \quad U(\tau, 1) = 0$$

   using the steps given above.

   Hint: $\cos(n\pi x)$ is an orthogonal set on $(0, 1)$. *The result will be a Fourier cosine series representation of 1.*

3. Plot U versus ξ for $\tau = 0.001$, 0.01, and 0.1 in Eq. (2.18). Comment.

Example 2.2. A Steady Heat Transfer Problem in Two Dimensions

Heat is conducted in a region of height a and width b. Temperature is a function of two space dimensions and independent of time. Three sides are at temperature u_0 and the fourth side is at temperature u_1. The formulation is as follows:

$$\frac{\partial^2 u}{\partial x^2} + \frac{\partial^2 u}{\partial y^2} = 0 \qquad (2.19)$$

with boundary conditions

$$u(0, x) = u(b, x) = u(y, a) = u_0$$
$$u(y, 0) = u_1 \qquad (2.20)$$

2.1.6 Scales and Dimensionless Variables

First note that there are two obvious length scales, a and b. We can choose either one of them to nondimensionalize x and y. We define

$$\xi = x/a \quad \text{and} \quad \eta = y/b \qquad (2.21)$$

so that both dimensionless lengths are normalized.

To normalize temperature we choose

$$U = \frac{u - u_0}{u_1 - u_0} \tag{2.22}$$

The problem statement reduces to

$$U_{\xi\xi} + \left(\frac{a}{b}\right)^2 U_{\eta\eta} = 0 \tag{2.23}$$

$$U(0, \xi) = U(1, \xi) = U(\eta, 1) = 0$$

$$U(\eta, 0) = 1 \tag{2.24}$$

2.1.7   Separation of Variables

As before, we assume a solution of the form $U(\xi, n) = X(\xi)Y(\eta)$. We substitute this into the differential equation and obtain

$$Y(\eta)X_{\xi\xi}(\xi) = -X(\xi)\left(\frac{a}{b}\right)^2 Y_{\eta\eta}(\eta) \tag{2.25}$$

Next we divide both sides by $U(\xi, n)$ and obtain

$$\frac{X_{\xi\xi}}{X} = -\left(\frac{a}{b}\right)^2 \frac{Y_{nn}}{Y} = \pm\lambda^2 \tag{2.26}$$

In order for the function only of ξ on the left-hand side of this equation to be equal to the function only of η on the right-hand side, both must be constant.

2.1.8   Choosing the Sign of the Separation Constant

However in this case it is not as clear as the case of Example 1 what the sign of this constant must be. Hence we have designated the constant as $\pm\lambda^2$ so that for real values of λ the $\pm$ sign determines the sign of the constant. Let us proceed by choosing the negative sign and see where this leads.

Thus

$$X_{\xi\xi} = -\lambda^2 X$$

$$Y(\eta)X(0) = 1$$

$$Y(\eta)X(1) = 0 \tag{2.27}$$

or

$$X(0) = 1$$

$$X(1) = 0 \tag{2.28}$$

and

$$Y_{\eta\eta} = \mp \left(\frac{b}{a}\right)^2 \lambda^2 Y \qquad (2.29)$$

$$X(\xi)Y(0) = X(\xi)Y(1) = 0$$

$$Y(0) = Y(1) = 0 \qquad (2.30)$$

The solution of the differential equation in the η direction is

$$Y(\eta) = A \cosh(b\lambda\eta/a) + B \sinh(b\lambda\eta/a) \qquad (2.31)$$

Applying the first boundary condition (at $\eta = 0$) we find that $A = 0$. When we apply the boundary condition at $\eta = 1$ however, we find that it requires that

$$0 = B \sinh(b\lambda/a) \qquad (2.32)$$

so that either $B = 0$ or $\lambda = 0$. Neither of these is acceptable since either would require that $Y(\eta) = 0$ for all values of η.

We next try the positive sign. In this case

$$X_{\xi\xi} = \lambda^2 X \qquad (2.33)$$

$$Y_{\eta\eta} = -\left(\frac{b}{a}\right)^2 \lambda^2 Y \qquad (2.34)$$

with the same boundary conditions given above. The solution for $Y(\eta)$ is now

$$Y(\eta) = A \cos(b\lambda\eta/a) + B \sin(b\lambda\eta/a) \qquad (2.35)$$

The boundary condition at $\eta = 0$ requires that

$$0 = A \cos(0) + B \sin(0) \qquad (2.36)$$

so that again $A = 0$. The boundary condition at $\eta = 1$ requires that

$$0 = B \sin(b\lambda/a) \qquad (2.37)$$

Since we don't want B to be zero, we can satisfy this condition if

$$\lambda_n = an\pi/b, \quad n = 0, 1, 2, 3, \ldots \qquad (2.38)$$

Thus

$$Y(\eta) = B \sin(n\pi\eta) \qquad (2.39)$$

Solution for $X(\xi)$ yields hyperbolic functions.

$$X(\xi) = C \cosh(\lambda_n \xi) + D \sinh(\lambda_n \xi) \qquad (2.40)$$

The boundary condition at $\xi = 1$ requires that

$$0 = C \cosh(\lambda_n) + D \sinh(\lambda_n) \qquad (2.41)$$

or, solving for C in terms of D,

$$C = -D \tanh(\lambda_n) \qquad (2.42)$$

One solution of our problem is therefore

$$U_n(\xi, \eta) = K_n \sin(n\pi\eta)[\sinh(an\pi\xi/b) - \cosh(an\pi\xi/b) \tanh(an\pi/b)] \qquad (2.43)$$

2.1.9   Superposition

According to the superposition theorem (Theorem 2) we can now form a solution as

$$U(\xi, \eta) = \sum_{n=0}^{\infty} K_n \sin(n\pi\eta)[\sinh(an\pi\xi/b) - \cosh(an\pi\xi/b) \tanh(an\pi/b)] \qquad (2.44)$$

The final boundary condition (the nonhomogeneous one) can now be applied,

$$1 = -\sum_{n=1}^{\infty} K_n \sin(n\pi\eta) \tanh(an\pi/b) \qquad (2.45)$$

2.1.10   Orthogonality

We have already noted that the sine function is an orthogonal function as defined on $(0, 1)$. Thus, we multiply both sides of this equation by $\sin(m\pi\eta)d\eta$ and integrate over $(0, 1)$, noting that according to the orthogonality theorem (Theorem 3) the integral is zero unless $n = m$. The result is

$$\int_{\eta=0}^{1} \sin(n\pi\eta)d\eta = -K_n \int_{\eta=0}^{1} \sin^2(n\pi\eta)d\eta \tanh(an\pi/b) \qquad (2.46)$$

$$\frac{1}{n\pi}[1 - (-1)^n] = -K_n \tanh(an\pi/b)\frac{1}{2} \qquad (2.47)$$

$$K_n = -\frac{2[1 - (-1)^n]}{n\pi \tanh(an\pi/b)} \qquad (2.48)$$

The solution is represented by the infinite series

$$U(\xi, \eta) = \sum_{n=1}^{\infty} \frac{2[1 - (-1)^n]}{n\pi \tanh(an\pi/b)} \sin(n\pi \eta)$$

$$\times [\cosh(an\pi\xi/b)\tanh(an\pi/b) - \sinh(an\pi\xi/b)] \qquad (2.49)$$

2.1.11 Lessons
The methodology for this problem is the same as in Example 1.

Example 2.3. A Steady Conduction Problem in Two Dimensions: Addition of Solutions

We now illustrate a problem in which two of the boundary conditions are nonhomogeneous. Since the problem and the boundary conditions are both linear we can simply break the problem into two problems and add them. Consider steady conduction in a square region L by L in size. Two sides are at temperature u_0 while the other two sides are at temperature u_1.

$$u_{xx} + u_{yy} = 0 \qquad (2.50)$$

We need four boundary conditions since the differential equation is of order 2 in both independent variables.

$$u(0, y) = u(L, y) = u_0 \qquad (2.51)$$
$$u(x, 0) = u(x, L) = u_1 \qquad (2.52)$$

2.1.12 Scales and Dimensionless Variables
The length scale is L, so we let $\xi = x/L$ and $\eta = y/L$. We can make the first two boundary conditions homogeneous while normalizing the second two by defining a dimensionless temperature as

$$U = \frac{u - u_0}{u_1 - u_0} \qquad (2.53)$$

Then

$$U_{\xi\xi} + U_{\eta\eta} = 0 \qquad (2.54)$$
$$U(0, \eta) = U(1, \eta) = 0 \qquad (2.55)$$
$$U(\xi, 0) = U(\xi, 1) = 1 \qquad (2.56)$$

2.1.13 Getting to One Nonhomogeneous Condition
There are two nonhomogeneous boundary conditions, so we must find a way to only have one. Let $U = V + W$ so that we have two problems, each with one nonhomogeneous boundary

condition.

$$W_{\xi\xi} + W_{\eta\eta} = 0 \qquad (2.57)$$

$$W(0, \eta) = W(1, \eta) = W(\xi, 0) = 0 \qquad (2.58)$$

$$W(\xi, 1) = 1$$

$$V_{\xi\xi} + V_{\eta\eta} = 0 \qquad (2.59)$$

$$V(0, \eta) = V(1, \eta) = V(\xi, 1) = 0 \qquad (2.60)$$

$$V(\xi, 0) = 1$$

(It should be clear that these two problems are identical if we put $V = W(1 - \eta)$. We will therefore only need to solve for W.)

2.1.14 Separation of Variables

Separate variables by letting $W(\xi, \eta) = P(\xi)Q(\eta)$.

$$\frac{P_{\xi\xi}}{P} = -\frac{Q_{\eta\eta}}{Q} = \pm\lambda^2 \qquad (2.61)$$

2.1.15 Choosing the Sign of the Separation Constant

Once again it is not immediately clear whether to choose the plus sign or the minus sign. Let's see what happens if we choose the plus sign.

$$P_{\xi\xi} = \lambda^2 P \qquad (2.62)$$

The solution is exponentials or hyperbolic functions.

$$P = A\sinh(\lambda\xi) + B\cosh(\lambda\xi) \qquad (2.63)$$

Applying the boundary condition on $\xi = 0$, we find that $B = 0$. The boundary condition on $\xi = 1$ requires that $A\sinh(\lambda) = 0$, which can only be satisfied if $A = 0$ or $\lambda = 0$, which yields a trivial solution, $W = 0$, and is unacceptable. The only hope for a solution is thus choosing the minus sign.

If we choose the minus sign in Eq. (2.61) then

$$P_{\xi\xi} = -\lambda^2 P \qquad (2.64)$$

$$Q_{\eta\eta} = \lambda^2 Q \qquad (2.65)$$

with solutions

$$P = A\sin(\lambda\xi) + B\cos(\lambda\xi) \qquad (2.66)$$

and

$$Q = C \sinh(\lambda \eta) + D \cosh(\lambda \eta) \tag{2.67}$$

respectively. Remembering to *apply the homogeneous boundary conditions first*, we find that for $W(0, \eta) = 0$, $B = 0$ and for $W(1, \eta) = 0$, $\sin(\lambda) = 0$. Thus, $\lambda = n\pi$, our eigenvalues corresponding to the eigenfunctions $\sin(n\pi\xi)$. The last homogeneous boundary condition is $W(\xi, 0) = 0$, which requires that $D = 0$. There are an infinite number of solutions of the form

$$PQ_n = K_n \sinh(n\pi \eta) \sin(n\pi \xi) \tag{2.68}$$

2.1.16 Superposition
Since our problem is linear we apply superposition.

$$W = \sum_{n=1}^{\infty} K_n \sinh(n\pi \eta) \sin(n\pi \xi) \tag{2.69}$$

Applying the final boundary condition, $W(\xi, 1) = 1$

$$1 = \sum_{n=1}^{\infty} K_n \sinh(n\pi) \sin(n\pi \xi). \tag{2.70}$$

2.1.17 Orthogonality
Multiplying both sides of Eq. (2.70) by $\sin(m\pi\xi)$ and integrating over the interval $(0, 1)$

$$\int_0^1 \sin(m\pi\xi)d\xi = \sum_{n=0}^{\infty} K_n \sinh(n\pi) \int_0^1 \sin(n\pi \xi) \sin(m\pi \xi)d\xi \tag{2.71}$$

The orthogonality property of the sine eigenfunction states that

$$\int_0^1 \sin(n\pi \xi) \sin(m\pi \xi)d\xi = \begin{array}{l} 0, \quad m \neq n \\ 1/2, \ m = n \end{array} \tag{2.72}$$

Thus,

$$K_n = 2/\sinh(n\pi) \tag{2.73}$$

and

$$W = \sum_{n=0}^{\infty} \frac{2}{\sinh(n\pi)} \sinh(n\pi \eta) \sin(n\pi \xi) \tag{2.74}$$

Recall that

$$V = W(\xi, 1 - \eta) \quad \text{and} \quad U = V + W$$

2.1.18 Lessons

If there are two nonhomogeneous boundary conditions break the problem into two problems that can be added (since the equations are linear) to give the complete solution. If you are unsure of the sign of the separation constant just assume a sign and move on. *Listen to what the mathematics is telling you.* It will always tell you if you choose wrong.

Example 2.4. A Non-homogeneous Heat Conduction Problem

Consider now the arrangement above, but with a heat source, and with both boundaries held at the initial temperature u_0. The heat source is initially zero and is turned on at $t = 0^+$. *The exercise illustrates the method of solving the problem when the single nonhomogeneous condition is in the partial differential equation rather than one of the boundary conditions.*

$$\rho c\, u_t = k u_{xx} + q \tag{2.75}$$

$$u(0, x) = u_0$$

$$u(t, 0) = u_0 \tag{2.76}$$

$$u(t, L) = u_0$$

2.1.19 Scales and Dimensionless Variables

Observe that the length scale is still L, so we define $\xi = x/L$. Recall that $k/\rho c = \alpha$ is the diffusivity. How shall we nondimensionalize temperature? We want as many ones and zeros in coefficients in the partial differential equation and the boundary conditions as possible. Define $U = (u - u_0)/S$, where S stands for "something with dimensions of temperature" that we must find. Dividing both sides of the partial differential equation by q and substituting for x

$$\frac{L^2 S \rho c\, U_t}{q} = \frac{k S U_{\xi\xi}}{q} + 1 \tag{2.77}$$

Letting $S = q/k$ leads to one as the coefficient of the first term on the right-hand side. Choosing the same dimensionless time as before, $\tau = \alpha t / L^2$ results in one as the coefficient of the time derivative term. We now have

$$U_\tau = U_{\xi\xi} + 1 \qquad (2.78)$$

$$U(0, \xi) = 0$$

$$U(\tau, 0) = 0 \qquad (2.79)$$

$$U(\tau, 1) = 0$$

2.1.20 Relocating the Nonhomogeneity

We have only one nonhomogeneous condition, but it's in the wrong place. The differential equation won't separate. For example if we let $U(\xi, \tau) = P(\xi)G(\tau)$ and insert this into the partial differential equation and divide by PG, we find

$$\frac{G'(\tau)}{G} = \frac{P''(\xi)}{P} + \frac{1}{PG} \qquad (2.80)$$

The technique to deal with this is to relocate the nonhomogenous condition to the initial condition. Assume a solution in the form $U = W(\xi) + V(\tau, \xi)$. We now have

$$V_\tau = V_{\xi\xi} + W_{\xi\xi} + 1 \qquad (2.81)$$

If we set $W_{\xi\xi} = -1$, the differential equation for V becomes homogeneous. We then set both W and V equal to zero at $\xi = 0$ and 1 and $V(0, \xi) = -W(\xi)$

$$W_{\xi\xi} = -1 \qquad (2.82)$$

$$W(0) = W(1) = 0 \qquad (2.83)$$

and

$$V_\tau = V_{\xi\xi} \qquad (2.84)$$

$$V(0, \xi) = -W(\xi)$$

$$V(\tau, 0) = 0 \qquad (2.85)$$

$$V(\tau, 1) = 0$$

The solution for W is parabolic

$$W = \frac{1}{2}\xi(1 - \xi) \qquad (2.86)$$

2.1.21 Separating Variables

We now solve for V using separation of variables.

$$V = P(\tau)Q(\xi) \tag{2.87}$$

$$\frac{P_\tau}{P} = \frac{Q_{\xi\xi}}{Q} = \pm\lambda^2 \tag{2.88}$$

We must choose the minus sign once again (see Problem 1 above) to have a negative exponential for $P(\tau)$. (We will see later that it's not always so obvious.) $P = \exp(-\lambda^2\tau)$.

The solution for Q is once again sines and cosines.

$$Q = A\cos(\lambda\xi) + B\sin(\lambda\xi) \tag{2.89}$$

The boundary condition $V(\tau, 0) = 0$ requires that $Q(0) = 0$. Hence, $A = 0$. The boundary condition $V(\tau, 1) = 0$ requires that $Q(1) = 0$. Since B cannot be zero, $\sin(\lambda) = 0$ so that our eigenvalues are $\lambda = n\pi$ and our eigenfunctions are $\sin(n\pi\xi)$.

2.1.22 Superposition

Once again using linear superposition,

$$V = \sum_{n=0}^{\infty} B_n \exp(-n^2\pi^2\tau)\sin(n\pi\xi) \tag{2.90}$$

Applying the initial condition

$$\frac{1}{2}\xi(\xi - 1) = \sum_{n=1}^{\infty} B_n \sin(n\pi\xi) \tag{2.91}$$

This is a Fourier sine series representation of $\frac{1}{2}\xi(\xi - 1)$. We now use the orthogonality of the sine function to obtain the coefficients B_n.

2.1.23 Orthogonality

Using the concept of orthogonality again, we multiply both sides by $\sin(m\pi\xi)d\xi$ and integrate over the space noting that the integral is zero if m is not equal to n. Thus, since

$$\int_0^1 \sin^2(n\pi\xi)d\xi = \frac{1}{2} \tag{2.92}$$

$$B_n = \int_0^1 \xi(\xi - 1)\sin(n\pi\xi)d\xi \tag{2.93}$$

2.1.24 Lessons

When the differential equation is nonhomogeneous use the linearity of the differential equation to transfer the nonhomogeneous condition to one of the boundary conditions. Usually this will result in a homogeneous partial differential equation and an ordinary differential equation.

We pause here to note that while the method of separation of variables is straightforward in principle, a certain amount of intuition or, if you wish, cleverness is often required in order to put the equation and boundary conditions in an appropriate form. The student working diligently will soon develop these skills.

Problems

1. Using these ideas obtain a series solution to the boundary value problem

$$u_t = u_{xx}$$
$$u(t, 1) = 0$$
$$u(t, 0) = 0$$
$$u(0, x) = 1$$

2. Find a series solution to the boundary value problem

$$u_t = u_{xx} + x$$
$$u_x(t, 0) = 0$$
$$u(t, 1) = 0$$
$$u(0, x) = 0$$

2.2   VIBRATIONS

In vibrations problems the dependent variable occurs in the differential equation as a second-order derivative of the independent variable t. The methodology is, however, essentially the same as it is in the diffusion equation. We first apply separation of variables, then use the boundary conditions to obtain eigenfunctions and eigenvalues, and use the linearity and orthogonality principles and the single nonhomogeneous condition to obtain a series solution. Once again, if there are more than one nonhomogeneous condition we use the linear superposition principle to obtain solutions for each nonhomogeneous condition and add the resulting solutions. We illustrate these ideas with several examples.

Example 2.5. A Vibrating String

Consider a string of length L fixed at the ends. The string is initially held in a fixed position $y(0, x) = f(x)$, where it is clear that $f(x)$ must be zero at both $x = 0$ and $x = L$. The boundary value problem is as follows:

$$y_{tt} = a^2 y_{xx} \tag{2.94}$$
$$y(t, 0) = 0$$
$$y(t, L) = 0 \tag{2.95}$$
$$y(0, x) = f(x)$$
$$y_t(0, x) = 0$$

2.2.1 Scales and Dimensionless Variables

The problem has the obvious length scale L. Hence let $\xi = x/L$. Now let $\tau = ta/L$ and the equation becomes

$$y_{\tau\tau} = y_{\xi\xi} \tag{2.96}$$

One could now nondimensionalize y, for example, by defining a new variable as $f(x)/f_{max}$, but it wouldn't simplify things. The boundary conditions remain the same except t and x are replaced by τ and ξ.

2.2.2 Separation of Variables

You know the dance. Let $y = P(\tau)Q(\xi)$. Differentiating and substituting into Eq. (2.96),

$$P_{\tau\tau}Q = PQ_{\xi\xi} \tag{2.97}$$

Dividing by PQ and noting that $P_{\tau\tau}/P$ and $Q_{\xi\xi}/Q$ cannot be equal to one another unless they are both constants, we find

$$P_{\tau\tau}/P = Q_{\xi\xi}/Q = \pm\lambda^2 \tag{2.98}$$

It should be physically clear that we want the minus sign. Otherwise both solutions will be hyperbolic functions. However if you choose the plus sign you will immediately find that the boundary conditions on ξ cannot be satisfied. Refer back to (2.63) and the sentences following.

The two ordinary differential equations and homogeneous boundary conditions are

$$P_{\tau\tau} + \lambda^2 P = 0 \tag{2.99}$$
$$P_\tau(0) = 0$$

and

$$Q_{\xi\xi} + \lambda^2 Q = 0 \tag{2.100}$$
$$Q(0) = 0$$
$$Q(1) = 0$$

The solutions are

$$P = A\sin(\lambda\tau) + B\cos(\lambda\tau) \tag{2.101}$$
$$Q = C\sin(\lambda\xi) + D\cos(\lambda\xi) \tag{2.102}$$

The first boundary condition of Eq. (2.100) requires that $D = 0$. The second requires that $C\sin(\lambda)$ be zero. Our eigenvalues are again $\lambda_n = n\pi$. The boundary condition at $\tau = 0$, that $P_\tau = 0$ requires that $A = 0$. Thus

$$PQ_n = K_n\sin(n\pi\xi)\cos(n\pi\tau) \tag{2.103}$$

The final form of the solution is then

$$y(\tau, \xi) = \sum_{n=0}^{\infty} K_n\sin(n\pi\xi)\cos(n\pi\tau) \tag{2.104}$$

2.2.3 Orthogonality
Applying the final (nonhomogeneous) boundary condition (the initial position).

$$f(\xi) = \sum_{n=0}^{\infty} K_n\sin(n\pi\xi) \tag{2.105}$$

In particular, if $f(x) = hx, \quad 0 < x < 1/2$

$$= h(1 - x), \qquad 1/2 < x < 1 \tag{2.106}$$

$$\int_0^1 f(x)\sin(n\pi x)dx = \int_0^{1/2} hx\sin(n\pi x)dx + \int_{1/2}^1 h(1 - x)\sin(n\pi x)dx$$

$$= \frac{2h}{n^2\pi^2}\sin\left(\frac{n\pi}{2}\right) = \frac{2h}{n^2\pi^2}(-1)^{n+1} \tag{2.107}$$

and

$$\int_0^1 K_n\sin^2(n\pi x)dx = K_n/2 \tag{2.108}$$

so that

$$y = \frac{4h}{\pi^2} \sum_{n=1}^{\infty} \frac{(-1)^{n+1}}{n^2} \sin(n\pi\xi) \cos(n\pi\tau) \qquad (2.109)$$

2.2.4   Lessons

The solutions are in the form of infinite series. The coefficients of the terms of the series are determined by using the fact that the solutions of at least one of the ordinary differential equations are orthogonal functions. The orthogonality condition allows us to calculate these coefficients.

Problem

1. Solve the boundary value problem

$$u_{tt} = u_{xx}$$

$$u(t, 0) = u(t, 1) = 0$$

$$u(0, x) = 0$$

$$u_t(0, x) = f(x)$$

Find the special case when $f(x) = \sin(\pi x)$.

FURTHER READING

V. Arpaci, *Conduction Heat Transfer*. Reading, MA: Addison-Wesley, 1966.

J. W. Brown and R. V. Churchill, *Fourier Series and Boundary Value Problems*. 6th edition. New York: McGraw-Hill, 2001.

CHAPTER 3

Orthogonal Sets of Functions

In this chapter we elaborate on the concepts of orthogonality and Fourier series. We begin with the familiar concept of orthogonality of vectors. We then extend the idea to orthogonality of functions and the use of this idea to represent general functions as Fourier series—series of orthogonal functions.

Next we show that solutions of a fairly general linear ordinary differential equation—the Sturm–Liouville equation—are orthogonal functions. Several examples are given.

3.1 VECTORS

We begin our study of orthogonality with the familiar topic of orthogonal vectors. Suppose $\mathbf{u}(1)$, $\mathbf{u}(2)$, and $\mathbf{u}(3)$ are the three rectangular component vectors in an ordinary three-dimensional space. The norm of the vector (its length) $||\mathbf{u}||$ is

$$||\mathbf{u}|| = [u(1)^2 + u(2)^2 + u(3)^2]^{1/2} \tag{3.1}$$

If $||\mathbf{u}|| = 1$, $\mathbf{u}$ is said to be normalized. If $||\mathbf{u}|| = 0$, $\mathbf{u}(r) = 0$ for each r and $\mathbf{u}$ is the zero vector.

A linear combination of two vectors $\mathbf{u}_1$ and $\mathbf{u}_2$ is

$$\mathbf{u} = c_1\mathbf{u}_1 + c_2\mathbf{u}_2, \tag{3.2}$$

The scalar or inner product of the two vectors $\mathbf{u}_1$ and $\mathbf{u}_2$ is defined as

$$(\mathbf{u}_1, \mathbf{u}_2) = \sum_{r=1}^{3} u_1(r)u_2(r) = ||u_1||\,||u_2||\cos\theta \tag{3.3}$$

3.1.1 Orthogonality of Vectors

If neither $\mathbf{u}_1$ nor $\mathbf{u}_2$ is the zero vector and if

$$(\mathbf{u}_1, \mathbf{u}_2) = 0 \tag{3.4}$$

then $\theta = \pi/2$ and the vectors are *orthogonal*. The norm of a vector $\mathbf{u}$ is

$$||\mathbf{u}|| = (\mathbf{u}, \mathbf{u})^{1/2} \tag{3.5}$$

3.1.2   Orthonormal Sets of Vectors

The vector $\Phi_n = \mathbf{u}_n/\|\mathbf{u}_n\|$ has magnitude unity, and if $\mathbf{u}_1$ and $\mathbf{u}_2$ are orthogonal then Φ_1 and Φ_2 are orthonormal and their inner product is

$$(\Phi_n, \Phi_m) = \delta_{nm} = 0, \; m \neq n \qquad (3.6)$$

$$= 1, \; m = n$$

where δ_{nm} is called the Kronecker delta.

If Φ_1, Φ_2, and Φ_3 are three vectors that are mutually orthogonal to each other then every vector in three-dimensional space can be written as a linear combination of Φ_1, Φ_2, and Φ_3; that is,

$$\mathbf{f}(r) = c_1\Phi_1 + c_2\Phi_2 + c_3\Phi_3 \qquad (3.7)$$

Note that due to the fact that the vectors Φ_n form an orthonormal set,

$$(\mathbf{f}, \Phi_1) = c_1, (\mathbf{f}, \Phi_2) = c_2, (\mathbf{f}, \Phi_3) = c_3 \qquad (3.8)$$

Simply put, suppose the vector $\mathbf{f}$ is

$$\mathbf{f} = 2\Phi_1 + 4\Phi_2 + \Phi_3. \qquad (3.9)$$

Taking the inner product of $\mathbf{f}$ with Φ_1 we find that

$$(\mathbf{f}, \Phi_1) = 2(\Phi_1, \Phi_1) + 4(\Phi_1, \Phi_2) + (\Phi_1, \Phi_3) \qquad (3.10)$$

and according to Eq. (3.8) $c_1 = 2$. Similarly, $c_2 = 4$ and $c_3 = 1$.

3.2   FUNCTIONS

3.2.1   Orthonormal Sets of Functions and Fourier Series

Suppose there is a set *of orthonormal functions* $\Phi_n(x)$ defined on an interval $a < x < b$ ($\sqrt{2}\sin(n\pi x)$ on the interval $0 < x < 1$ is an example). A set of orthonormal *functions* is defined as one whose inner product, defined as $\int_{x=a}^{b} \Phi_n(x)\Phi_m(x)dx$, is

$$(\Phi_n, \Phi_m) = \int_{x=a}^{b} \Phi_n \Phi_m \, dx = \delta_{nm} \qquad (3.11)$$

Suppose we can express a function as an infinite series of these orthonormal functions,

$$f(x) = \sum_{n=0}^{\infty} c_n \Phi_n \quad \text{on} \quad a < x < b \qquad (3.12)$$

Equation (3.12) is called a *Fourier series* of $f(x)$ in terms of the orthonormal function set $\Phi_n(x)$.

If we now form the inner product of Φ_m with both sides of Eq. (3.12) and use the definition of an orthonormal function set as stated in Eq. (3.11) we see that the inner product of $f(x)$ and $\Phi_n(x)$ is c_n.

$$c_n \int_{x=a}^{b} \Phi_n^2(\xi)d\xi = c_n = \int_{x=a}^{b} f(\xi)\,\Phi_n(\xi)d\xi \qquad (3.13)$$

In particular, consider a set of functions Ψ_n that are orthogonal on the interval (a, b) so that

$$\int_{x=a}^{b} \Psi_n(\xi)\Psi_m(\xi)d\xi = 0, \quad m \neq n$$
$$= \|\Psi_n\|^2, \quad m = n \qquad (3.14)$$

where $\|\Psi_n\|^2 = \int_{x=a}^{b} \Psi_n^2(\xi)d\xi$ is called the square of the norm of Ψ_n. The functions

$$\frac{\Psi_n}{\|\Psi_n\|} = \Phi_n \qquad (3.15)$$

then form an orthonormal set. We now show how to form the series representation of the function $f(x)$ as a series expansion in terms of the orthogonal (but not orthonormal) set of functions $\Psi_n(x)$.

$$f(x) = \sum_{n=0}^{\infty} \frac{\Psi_n}{\|\Psi_n\|} \int_{\xi=a}^{b} f(\xi)\frac{\Psi_n(\xi)}{\|\Psi_n\|}d\xi = \sum_{n=0}^{\infty} \Psi_n \int_{\xi=a}^{b} f(\xi)\frac{\Psi_n(\xi)}{\|\Psi_n\|^2}d\xi \qquad (3.16)$$

This is called a Fourier series representation of the function $f(x)$.

As a concrete example, the square of the norm of the sine function on the interval $(0, \pi)$ is

$$\| \sin(nx)\|^2 = \int_{\xi=0}^{\pi} \sin^2(n\xi)d\xi = \frac{\pi}{2} \qquad (3.17)$$

so that the corresponding orthonormal function is

$$\Phi = \sqrt{\frac{2}{\pi}} \sin(nx) \qquad (3.18)$$

A function can be represented by a series of sine functions on the interval $(0, \pi)$ as

$$f(x) = \sum_{n=0}^{\infty} \sin(nx) \int_{\varsigma=0}^{\pi} \frac{\sin(n\varsigma)}{\pi/2}f(\varsigma)d\varsigma \qquad (3.19)$$

This is a *Fourier sine series*.

3.2.2 Best Approximation

We next ask whether, since we can never sum to infinity, the values of the constants c_n in Eq. (3.13) give the most accurate approximation of the function. To illustrate the idea we return to the idea of *orthogonal vectors* in three-dimensional space. Suppose we want to approximate a three-dimensional vector with a two-dimensional vector. What will be the components of the two-dimensional vector that best approximate the three-dimensional vector?

Let the three-dimensional vector be $\mathbf{f} = c_1\boldsymbol{\Phi}_1 + c_2\boldsymbol{\Phi}_2 + c_3\boldsymbol{\Phi}_3$. Let the two-dimensional vector be $\mathbf{k} = a_1\boldsymbol{\Phi}_1 + a_2\boldsymbol{\Phi}_2$. We wish to minimize $||\mathbf{k} - \mathbf{f}||$.

$$||\mathbf{k} - \mathbf{f}|| = \left\{(a_1 - c_1)^2 + (a_2 - c_2)^2 + c_3^2\right\}^{1/2} \tag{3.20}$$

It is clear from the above equation (and also from Fig. 3.1) that this will be minimized when $a_1 = c_1$ and $a_2 = c_2$.

Turning now to the *orthogonal function* series, we attempt to minimize the difference between the function with an infinite number of terms and the summation only to some finite value m. The square of the error is

$$E^2 = \int_{x=a}^{b} (f(x) - K_m(x))^2 dx = \int_{x=a}^{b} \left[f^2(x) + K^2(x) - 2f(x)K(x)\right]dx \tag{3.21}$$

where

$$f(x) = \sum_{n=1}^{\infty} c_n\Phi_n(x) \tag{3.22}$$

and

$$K_m = \sum_{n=1}^{m} a_n\Phi_n(x) \tag{3.23}$$

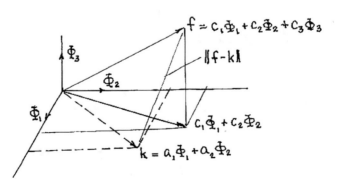

FIGURE 3.1: Best approximation of a three-dimensional vector in two dimensions

Noting that

$$\int_{x=a}^{b} K_m^2(x)dx = \sum_{n=1}^{m}\sum_{j=1}^{m} a_n a_j \int_{x=a}^{b} \Phi_n(x)\Phi_j(x)dx = \sum_{n=1}^{m} a_n^2 = a_1^2 + a_2^2 + a_3^2 + \cdots + a_m^2$$

(3.24)

and

$$\int_{x=a}^{b} f(x)K(x)dx = \sum_{n=1}^{\infty}\sum_{j=1}^{m} c_n a_j \int_{x=a}^{b} \Phi_n(x)\Phi_j(x)dx$$

$$= \sum_{n=1}^{m} c_n a_n = c_1 a_1 + c_2 a_2 + \cdots + c_m a_m$$

(3.25)

$$E^2 = \int_{x=a}^{b} f^2(x)dx + a_1^2 + \cdots + a_m^2 - 2a_1 c_1 - \cdots - 2a_m c_m$$

(3.26)

Now add and subtract $c_1^2, c_2^2, \ldots, c_m^2$. Thus Eq. (3.26) becomes

$$E^2 = \int_{x=a}^{b} f^2(x)dx - c_1^2 - c_2^2 - \cdots - c_m^2 + (a_1 - c_1)^2 + (a_2 - c_2)^2 + \cdots + (a_m - c_m)^2$$

(3.27)

which is clearly minimized when $a_n = c_n$.

3.2.3   Convergence of Fourier Series

We briefly consider the question of whether the Fourier series actually converges to the function $f(x)$ for all values, say, on the interval $a \le x \le b$. The series will converge to the function if the value of E defined in (3.19) approaches zero as m approaches infinity. Suffice to say that this is true for functions that are continuous with piecewise continuous first derivatives, that is, most physically realistic temperature distributions, displacements of vibrating strings and bars. In each particular situation, however, one should use the various convergence theorems that are presented in most elementary calculus books. Uniform convergence of Fourier series is discussed extensively in the book *Fourier Series and Boundary Value Problems* by James Ward Brown and R. V. Churchill. In this chapter we give only a few physically clear examples.

3.2.4   Examples of Fourier Series

Example 3.1.   Determine a Fourier sine series representation of $f(x) = x$ on the interval $(0, 1)$. The series will take the form

$$x = \sum_{j=0}^{\infty} c_j \sin(j\pi x) \qquad (3.28)$$

since the $\sin(j\pi x)$ forms an orthogonal set on $(0, 1)$, multiply both sides by $\sin(k\pi x)dx$ and integrate over the interval on which the function is orthogonal.

$$\int_{x=0}^{1} x \sin(k\pi x)dx = \sum_{k=0}^{\infty} \int_{x=0}^{1} c_j \sin(j\pi x) \sin(k\pi x)dx \qquad (3.29)$$

Noting that all of the terms on the right-hand side of (2.20) are zero except the one for which $k = j$,

$$\int_{x=0}^{1} x \sin(j\pi x)dx = c_j \int_{x=0}^{1} \sin^2(j\pi x)dx \qquad (3.30)$$

After integrating we find

$$\frac{(-1)^{j+1}}{j\pi} = \frac{c_j}{2} \qquad (3.31)$$

Thus,

$$x = \sum_{j=0}^{\infty} \frac{(-1)^{j+1}}{j\pi} 2 \sin(j\pi x) \qquad (3.32)$$

This is an alternating sign series in which the coefficients always decrease as j increases, and it therefore converges. The sine function is periodic and so the series must also be a periodic function beyond the interval $(0, 1)$. The series outside this interval forms the *periodic continuation* of the series. Note that the sine is an odd function so that $\sin(j\pi x) = -\sin(-j\pi x)$. Thus the periodic continuation looks like Fig. 3.2. The series converges everywhere, but at $x = 1$ it is identically zero instead of one. It converges to $1 - \varepsilon$ arbitrarily close to $x = 1$.

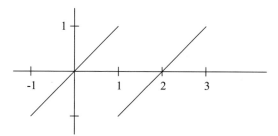

FIGURE 3.2: The periodic continuation of the function x represented by the sine series

Example 3.2. Find a Fourier cosine for $f(x) = x$ on the interval $(0, 1)$. In this case

$$x = \sum_{n=0}^{\infty} c_n \cos(n\pi x) \tag{3.33}$$

Multiply both sides by $\cos(m\pi x)dx$ and integrate over $(0, 1)$.

$$\int_{x=0}^{1} x \cos(m\pi x)dx = \sum_{n=0}^{\infty} c_n \int_{x=0}^{1} \cos(m\pi x) \cos(n\pi x)dx \tag{3.34}$$

and noting that $\cos(n\pi x)$ is an orthogonal set on $(0, 1)$ all terms in (2.23) are zero except when $n = m$. Evaluating the integrals,

$$\frac{c_n}{2} = \frac{[(-1)^2 - 1]}{(n\pi)^2} \tag{3.35}$$

There is a problem when $n = 0$. Both the numerator and the denominator are zero there. However we can evaluate c_0 by noting that according to Eq. (3.26)

$$\int_{x=0}^{1} xdx = c_0 = \frac{1}{2} \tag{3.36}$$

and the cosine series is therefore

$$x = \frac{1}{2} + \sum_{n=1}^{\infty} 2\frac{[(-1)^n - 1]}{(n\pi)^2} \cos(n\pi x) \tag{3.37}$$

The series converges to x everywhere. Since $\cos(n\pi x) = \cos(-n\pi x)$ it is an even function and its *periodic continuation* is shown in Fig. 3.3. Note that the sine series is discontinuous at $x = 1$, while the cosine series is continuous everywhere. (Which is the better representation?)

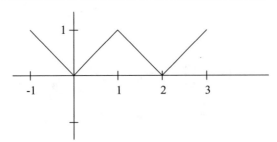

FIGURE 3.3: The periodic continuation of the series in Example 3.2

It should be clear from the above examples that in general a Fourier sine/cosine series of a function $f(x)$ defined on $0 \leq x \leq 1$ can be written as

$$f(x) = \frac{c_0}{2} + \sum_{n=1}^{\infty} c_n \cos(n\pi x) + \sum_{n=1}^{\infty} b_n \sin(n\pi x) \qquad (3.38)$$

where

$$c_n = \frac{\int_{x=0}^{1} f(x) \cos(n\pi x)dx}{\int_{x=0}^{1} \cos^2(n\pi x)dx} \qquad n = 0, 1, 2, 3, \ldots$$

$$b_n = \frac{\int_{x=0}^{1} f(x) \sin(n\pi x)dx}{\int_{x=0}^{1} \sin^2(n\pi x)dx} \qquad n = 1, 2, 3, \ldots \qquad (3.39)$$

Problems

1. Show that

$$\int_{x=0}^{\pi} \sin(nx) \sin(mx)dx = 0$$

when $n \neq m$.

2. Find the Fourier sine series for $f(x) = 1 - x$ on the interval $(0, 1)$. Sketch the periodic continuation. Sum the series for the first five terms and sketch over two periods. Discuss convergence of the series, paying special attention to convergence at $x = 0$ and $x = 1$.

3. Find the Fourier cosine series for $1 - x$ on $(0, 1)$. Sketch the periodic continuation. Sum the first two terms and sketch. Sum the first five terms and sketch over two periods. Discuss convergence, paying special attention to convergence at $x = 0$ and $x = 1$.

3.3   STURM–LIOUVILLE PROBLEMS: ORTHOGONAL FUNCTIONS

We now proceed to show that solutions of a certain ordinary differential equation with certain boundary conditions (called a Sturm–Liouville problem) are *orthogonal functions with respect to a weighting function*, and that therefore a well-behaved function can be represented by an infinite series of these orthogonal functions (called eigenfunctions), as in Eqs. (3.12) and (3.16).

Recall that the problem

$$X_{xx} + \lambda^2 X = 0, \; X(0) = 0, \; X(1) = 0 \qquad 0 \le x \le 1 \qquad (3.40)$$

has solutions only for $\lambda = n\pi$ and that the solutions, $\sin(n\pi x)$ are orthogonal on the interval (0, 1). The sine functions are called eigenfunctions and $\lambda = n\pi$ are called eigenvalues.

As another example, consider the problem

$$X_{xx} + \lambda^2 X = 0 \qquad (3.41)$$

with boundary conditions

$$X(0) = 0$$
$$X(1) + HX_x(1) = 0 \qquad (3.42)$$

The solution of the differential equation is

$$X = A\sin(\lambda x) + B\cos(\lambda x)) \qquad (3.43)$$

The first boundary condition guarantees that $B = 0$. The second boundary condition is satisfied by the equation

$$A[\sin(\lambda) + H\lambda \cos(\lambda)] = 0 \qquad (3.44)$$

Since A cannot be zero, this implies that

$$-\tan(\lambda) = H\lambda. \qquad (3.45)$$

The *eigenfunctions* are $\sin(\lambda x)$ and the *eigenvalues* are solutions of Eq. (3.45). This is illustrated graphically in Fig. 3.4.

We will generally be interested in the fairly general linear second-order differential equation and boundary conditions given in Eqs. (3.46) and (3.47).

$$\frac{d}{dx}\left[r(x)\frac{dX}{dx}\right] + [q(x) + \lambda p(x)]X = 0 \quad a \le x \le b \qquad (3.46)$$

$$a_1 X(a) + a_2 dX(a)/dx = 0$$
$$b_1 X(b) + b_2 dX(b)/dx = 0 \qquad (3.47)$$

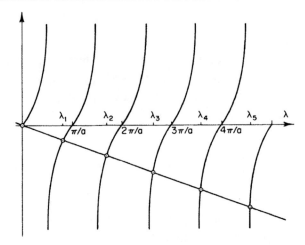

FIGURE 3.4: Eigenvalues of $-\tan(\lambda) = H\lambda$

Solutions exist only for discrete values λ_n the *eigenvalues*. The corresponding solutions $X_n(x)$ are the *eigenfunctions*.

3.3.1   Orthogonality of Eigenfunctions

Consider two solutions of (3.46) and (3.47), X_n and X_m corresponding to eigenvalues λ_n and λ_m. The primes denote differentiation with respect to x.

$$(r X_m')' + q X_m = -\lambda_m p X_m \qquad (3.48)$$

$$(r X_n')' + q X_n = -\lambda_n p X_n \qquad (3.49)$$

Multiply the first by X_n and the second by X_m and subtract, obtaining the following:

$$(r X_n X_m' - r X_m X_n')' = (\lambda_n - \lambda_m) p X_m X_n \qquad (3.50)$$

Integrating both sides

$$r(X_m' X_n - X_n' X_m)_a^b = (\lambda_n - \lambda_m) \int_a^b p(x) X_n X_m dx \qquad (3.51)$$

Inserting the boundary conditions into the left-hand side of (3.51)

$$X_m'(b)X_n(b) - X_m'(a)X_n(a) - X_n'(b)X_m(b) + X_n'(a)X_m(a)$$

$$= -\frac{b_1}{b_2}X_m(b)X_n(b) + \frac{a_1}{a_2}X_m(a)X_n(a) - \frac{a_1}{a_2}X_n(a)X_m(a) + \frac{b_1}{b_2}X_m(b)X_n(b) = 0 \qquad (3.52)$$

Thus

$$(\lambda_n - \lambda_m) \int_a^b p(x) X_n X_m dx = 0, \quad m \neq n \qquad (3.53)$$

Notice that X_m and X_n are *orthogonal with respect to the weighting function p(x) on the interval (a, b)*. Obvious examples are the sine and cosine functions.

Example 3.3. Example 2.1 in Chapter 2 is an example in which the *eigenfunctions* are $\sin(\lambda_n \xi)$ and the *eigenvalues* are $(2n - 1)\pi/2$.

Example 3.4. If the boundary conditions in Example 2.1 in Chapter 2 are changed to

$$\Phi'(0) = 0 \qquad \Phi(1) = 0 \qquad (3.54)$$

we note that the general solution of the differential equation is

$$\Phi(\xi) = A \cos(\lambda \xi) + B \sin(\lambda \xi) \qquad (3.55)$$

The boundary conditions require that $B = 0$ and $\cos(\lambda) = 0$. The values of λ can take on any of the values $\pi/2, 3\pi/2, 5\pi/2, \ldots, (2n - 1)\pi/2$. The *eigenfunctions* are $\cos(\lambda_n \xi)$ and the *eigenvalue* are $\lambda_n = (2n - 1)\pi/2$.

Example 3.5. Suppose the boundary conditions in the original problem (Example 1, Chapter 2) take on the more complicated form

$$\Phi(0) = 0 \qquad \Phi(1) + h\Phi'(1) = 0 \qquad (3.56)$$

The first boundary condition requires that $B = 0$. The second boundary conditions require that

$$\sin(\lambda_n) + h\lambda_n \cos(\lambda_n) = 0, \text{ or} \qquad (3.57)$$

$$\lambda_n = -\frac{1}{h} \tan(\lambda_n) \qquad (3.58)$$

which is a transcendental equation that must be solved for the *eigenvalues*. The *eigenfunctions* are, of course, $\sin(\lambda_n x)$.

Example 3.6. A Physical Example: Heat Conduction in Cylindrical Coordinates

The heat conduction equation in cylindrical coordinates is

$$\frac{\partial u}{\partial t} = \frac{\partial^2 u}{\partial r^2} + \frac{1}{r}\frac{\partial u}{\partial r} \qquad 0 < r < 1 \qquad (3.59)$$

with boundary conditions at $R = 0$ and $r = 1$ and initial condition $u(0, r) = f(r)$.

Separating variables as $u = R(r)T(t)$,

$$\frac{1}{T}\frac{dT}{dt} = \frac{1}{R}\frac{d^2R}{dr^2} + \frac{1}{rR}\frac{dR}{dr} = -\lambda^2 \qquad 0 \le r \le 1, \quad 0 \le t \tag{3.60}$$

(Why the minus sign?)

The equation for $R(r)$ is

$$(rR')' + \lambda^2 r\, R = 0, \tag{3.61}$$

which is a Sturm–Liouville equation *with weighting function r*. It is an eigenvalue problem with an infinite number of eigenfunctions corresponding to the eigenvalues λ_n. There will be two solutions $R_1(\lambda_n r)$ and $R_2(\lambda_n r)$ for each λ_n. The solutions are called Bessel functions, and they will be discussed in Chapter 4.

$$R_n(\lambda_n r) = A_n R_1(\lambda_n r) + B_n R_2(\lambda_n r) \tag{3.62}$$

The boundary conditions on r are used to determine a relation between the constants A and B. For solutions $R(\lambda_n r)$ and $R(\lambda_m r)$

$$\int_0^1 r R(\lambda_n r) R(\lambda_m r)dr = 0, \quad n \ne m \tag{3.63}$$

is the orthogonality condition.

The solution for $T(t)$ is the exponential $e^{-\lambda_n^2 t}$ for all n. Thus, the solution of (3.60), because of superposition, can be written as an infinite series in a form something like

$$u = \sum_{n=0}^{\infty} K_n e^{-\lambda_n^2 t} R(\lambda_n r) \tag{3.64}$$

and the orthogonality condition is used to find K_n as

$$K_n = \int_{r=0}^1 f(r) R(\lambda_n r) r\, dr \Big/ \int_{r=0}^1 f(r) R^2(\lambda_n r) r\, dr \tag{3.65}$$

Problems

1. For Example 2.1 in Chapter 2 with the new boundary conditions described in Example 3.2 above, find K_n and write the infinite series solution to the revised problem.

FURTHER READING

J. W. Brown and R. V. Churchill, *Fourier Series and Boundary Value Problems*, 6th edition. New York: McGraw-Hill, 2001.

P. V. O'Neil, *Advanced Engineering Mathematics*. 5th edition. Brooks/Cole Thompson, Pacific Grove, CA, 2003.

CHAPTER 4

Series Solutions of Ordinary Differential Equations

4.1 GENERAL SERIES SOLUTIONS

The purpose of this chapter is to present a method of obtaining solutions of linear second-order ordinary differential equations in the form of Taylor series'. The methodology is then used to obtain solutions of two special differential equations, Bessel's equation and Legendre's equation. Properties of the solutions—Bessel functions and Legendre functions—which are extensively used in solving problems in mathematical physics, are discussed briefly. Bessel functions are used in solving both diffusion and vibrations problems in cylindrical coordinates. The functions $R(\lambda_n r)$ in Example 3.4 at the end of Chapter 3 are called Bessel functions. Legendre functions are useful in solving problems in spherical coordinates. Associated Legendre functions, also useful in solving problems in spherical coordinates, are briefly discussed.

4.1.1 Definitions

In this chapter we will be concerned with linear second-order equations. A general case is

$$a(x)u'' + b(x)u' + c(x)u = f(x) \tag{4.1}$$

Division by $a(x)$ gives

$$u'' + p(x)u' + q(x)u = r(x) \tag{4.2}$$

Recall that if $r(x)$ is zero the equation is *homogeneous*. The solution can be written as the sum of a *homogeneous solution* $u_h(x)$ and a *particular solution* $u_p(x)$. If $r(x)$ is zero, $u_p = 0$. The nature of the solution and the solution method depend on the nature of the coefficients $p(x)$ and $q(x)$. If each of these functions can be expanded in a Taylor series about a point x_0 the point is said to be an *ordinary point* and the function is *analytic* at that point. If either of the coefficients is not analytic at x_0, the point is a *singular point*. If x_0 is a singular point and if $(x - x_0)p(x)$ and $(x - x_0)^2 q(x)$ are analytic, then the singularities are said to be *removable* and the singular point is a *regular singular point*. If this is not the case the singular point is *irregular*.

4.1.2   Ordinary Points and Series Solutions

If the point x_0 is an ordinary point the dependent variable has a solution in the neighborhood of x_0 of the form

$$u(x) = \sum_{n=0}^{\infty} c_n(x - x_0)^n \qquad (4.3)$$

We now illustrate the solution method with two examples.

Example 4.1.   Find a series solution in the form of Eq. (4.3) about the point $x = 0$ of the differential equation

$$u'' + x^2 u = 0 \qquad (4.4)$$

The point $x = 0$ is an ordinary point so at least near $x = 0$ there is a solution in the form of the above series. Differentiating (4.3) twice and inserting it into (4.4)

$$u' = \sum_{n=0}^{\infty} n c_n x^{n-1}$$

$$u'' = \sum_{n=0}^{\infty} n(n - 1) c_n x^{n-2}$$

$$\sum_{n=0}^{\infty} n(n - 1) c_n x^{n-2} + \sum_{n=0}^{\infty} x^{n+2} c_n = 0 \qquad (4.5)$$

Note that the first term in the u' series is zero while the first two terms in the u'' series are zero. We can shift the indices in both summations so that the power of x is the same in both series by setting $n - 2 = m$ in the first series.

$$\sum_{n=0}^{\infty} n(n - 1) c_n x^{n-2} = \sum_{m=-2}^{\infty} (m + 2)(m + 1) c_{m+2} x^m = \sum_{m=0}^{\infty} (m + 2)(m + 1) c_{m+2} x^m \qquad (4.6)$$

Noting that m is a "dummy variable" and that the first two terms in the series are zero the series can be written as

$$\sum_{n=0}^{\infty} (n + 2)(n + 1) c_{n+2} x^n \qquad (4.7)$$

In a similar way we can write the second term as

$$\sum_{n=0}^{\infty} c_n x^{n+2} = \sum_{n=2}^{\infty} c_{n-2} x^n \qquad (4.8)$$

We now have

$$\sum_{n=0}^{\infty}(n+2)(n+1)c_{n+2}x^n + \sum_{n=2}^{\infty}c_{n-2}x^n = 0 \tag{4.9}$$

which can be written as

$$2c_2 + 6c_3x + \sum_{n=2}^{\infty}[(n+2)(n+1)c_{n+2} + c_{n-2}]x^n = 0 \tag{4.10}$$

Each coefficient of x^n must be zero in order to satisfy Eq. (4.10). Thus c_2 and c_3 must be zero and

$$c_{n+2} = -c_{n-2}/(n+2)(n+1) \tag{4.11}$$

while c_0 and c_1 remain arbitrary.

Setting $n = 2$, we find that $c_4 = -c_0/12$ and setting $n = 3$, $c_5 = -c_1/20$. Since c_2 and c_3 are zero, so are c_6, c_7, c_{10}, c_{11}, etc. Also, $c_8 = -c_4/(8)(7) = c_0/(4)(3)(8)(7)$ and

$$c_9 = -c_5/(9)(8) = c_1/(5)(4)(9)(8).$$

The first few terms of the series are

$$u(x) = c_0(1 - x^4/12 + x^6/672 + \cdots) + c_1(1 - x^5/20 + x^9/1440 + \cdots) \tag{4.12}$$

The values of c_0 and c_1 may be found from appropriate boundary conditions. These are both alternating sign series with each term smaller than the previous term at least for $x \le 1$ and it is therefore convergent at least under these conditions.

The constants c_0 and c_1 can be determined from boundary conditions. For example if $u(0) = 0$, $c_0 + c_1 = 0$, so $c_1 = -c_0$. If $u(1) = 1$,

$$c_0[-1/12 + 1/20 + 1/672 - 1/1440 + \cdots] = 1$$

Example 4.2. Find a series solution in the form of Eq. (4.3) of the differential equation

$$u'' + xu' + u = x^2 \tag{4.13}$$

valid near $x = 0$.

Assuming a solution in the form of (4.3), differentiating and inserting into (4.13),

$$\sum_{n=0}^{\infty}(n-1)nc_nx^{n-2} + \sum_{n=0}^{\infty}nc_nx^n + \sum_{n=0}^{\infty}c_nx^n - x^2 = 0 \tag{4.14}$$

Shifting the indices as before

$$\sum_{n=0}^{\infty}(n+2)(n+1)c_{n+2}x^n + \sum_{n=0}^{\infty}nc_nx^n + \sum_{n=0}^{\infty}c_nx^n - x^2 = 0 \qquad (4.15)$$

Once again, each of the coefficients of x^n must be zero.

Setting $n = 0$, we see that

$$n = 0 : 2c_2 + c_0 = 0, \qquad c_2 = -c_0/2 \qquad\qquad (4.16)$$
$$n = 1 : 6c_3 + 2c_1 = 0, \qquad c_3 = -c_1/3$$
$$n = 2 : 12c_4 + 3c_2 - 1 = 0, \qquad c_4 = (1 + 3c_0/2)/12$$
$$n > 2 : c_{n+2} = \frac{c_n}{n+2}$$

The last of these is called a *recurrence formula*.

Thus,

$$\begin{aligned} u = {} & c_0(1 - x^2/2 + x^4/8 - x^6/(8)(6) + \cdots) \\ & + c_1(x - x^3/3 + x^5/(3)(5) - x^7/(3)(5)(7) + \cdots) \\ & + x^4(1/12 - x^2/(12)(6) + \cdots) \end{aligned} \qquad (4.17)$$

Note that the series on the third line of (4.17) is the *particular solution* of (4.13). The constants c_0 and c_1 are to be evaluated using the boundary conditions.

4.1.3   Lessons: Finding Series Solutions for Differential Equations with Ordinary Points

If x_0 is an ordinary point assume a solution in the form of Eq. (4.3) and substitute into the differential equation. Then equate the coefficients of equal powers of x. This will give a recurrence formula from which two series may be obtained in terms of two arbitrary constants. These may be evaluated by using the two boundary conditions.

Problems

1. The differential equation

$$u'' + xu' + xu = x$$

has ordinary points everywhere. Find a series solution near $x = 0$.

2. Find a series solution of the differential equation

$$u'' + (1 + x^2)u = x$$

near $x = 0$ and identify the particular solution.

3. The differential equation

$$(1 - x^2)u'' + u = 0$$

has singular points at $x = \pm 1$, but is analytic near $x = 0$. Find a series solution that is valid near $x = 0$ and discuss the radius of convergence.

4.1.4 Regular Singular Points and the Method of Frobenius

If x_0 is a singular point in (4.2) there may not be a power series solution of the form of Eq. (4.3). In such a case we proceed by assuming a solution of the form

$$u(x) = \sum_{n=0}^{\infty} c_n(x - x_0)^{n+r} \qquad (4.18)$$

in which $c_0 \neq 0$ and r is any constant, not necessarily an integer. This is called the method of Frobenius and the series is called a Frobenius series. The Frobenius series need not be a power series because r may be a fraction or even negative. Differentiating once

$$u' = \sum_{n=0}^{\infty} (n + r)c_n(x - x_0)^{n+r-1} \qquad (4.19)$$

and differentiating again

$$u'' = \sum_{n=0}^{\infty} (n + r - 1)(n + r)c_n(x - x_0)^{n+r-2} \qquad (4.20)$$

These are then substituted into the differential equation, shifting is done where required so that each term contains x raised to the power n, and the coefficients of x^n are each set equal to zero. The coefficient associated with the lowest power of x will be a quadratic equation that can be solved for the index r. It is called an *indicial equation*. There will therefore be two roots of this equation corresponding to two series solutions. The values of c_n are determined as above by a *recurrence equation* for each of the roots. Three possible cases are important: (a) the roots are distinct and do not differ by an integer, (b) the roots differ by an integer, and (c) the roots are coincident, i.e., repeated. We illustrate the method by a series of examples.

Example 4.3 (distinct roots). Solve the equation

$$x^2u'' + x(1/2 + 2x)u' + (x - 1/2)u = 0 \qquad (4.21)$$

The coefficient of the u' term is

$$p(x) = \frac{(1/2 + 2x)}{x} \qquad (4.22)$$

and the coefficient of the u'' term is

$$q(x) = \frac{(x - 1/2)}{x^2} \qquad (4.23)$$

Both have singularities at $x = 0$. However multiplying $p(x)$ by x and $q(x)$ by x^2 the singularities are removed. Thus $x = 0$ is a regular singular point. Assume a solution in the form of the Frobenius series: $u = \sum_{n=0}^{\infty} c_n x^{n+r}$, differentiate twice and substitute into (4.21) obtaining

$$\sum_{n=0}^{\infty} (n+r)(n+r-1)x^{n+1} + \sum_{n=0}^{\infty} \frac{1}{2}(n+r)c_n x^{n+r} + \sum_{n=0}^{\infty} 2(n+r)c_n x^{n+r+1}$$

$$+ \sum_{n=0}^{\infty} c_n x^{n+r+1} - \sum_{n=0}^{\infty} \frac{1}{2} c_n x^{n+r} = 0 \qquad (4.24)$$

The indices of the third and fourth summations are now shifted as in Example 4.1 and we find

$$\left[r(r-1) + \frac{1}{2}r - \frac{1}{2} \right] c_0 x^r + \sum_{n=1}^{\infty} \left[(n+r)(n+r-1) + \frac{1}{2}(n+r) - \frac{1}{2} \right] c_n x^{n+r}$$

$$+ \sum_{n=1}^{\infty} [2(n+r-1) + 1] c_{n-1} x^{n+r} = 0 \qquad (4.25)$$

Each coefficient must be zero for the equation to be true. Thus the coefficient of the c_0 term must be zero since c_0 itself cannot be zero. This gives a quadratic equation to be solved for r, and this is called an *indicial equation* (since we are solving for the index, r).

$$r(r-1) + \frac{1}{2}r - \frac{1}{2} = 0 \qquad (4.26)$$

with $r = 1$ and $r = -1/2$. The coefficients of x^{n+r} must also be zero. Thus

$$[(n+r)(n+r-1) + 1/2(n+r) - 1/2]c_n + [2(n+r-1) + 1]c_{n-1} = 0 . \qquad (4.27)$$

The *recurrence equation* is therefore

$$c_n = -\frac{2(n+r-1) + 1}{(n+r)(n+r-1) + \frac{1}{2}(n+r) - \frac{1}{2}} c_{n-1} \qquad (4.28)$$

For the case of $r = 1$

$$c_n = -\frac{2n+1}{n\left(n + \frac{3}{2}\right)} c_{n-1} \qquad (4.29)$$

Computing a few of the coefficients,

$$c_1 = -\frac{3}{\frac{5}{2}}c_0 = -\frac{6}{5}c_0$$

$$c_2 = -\frac{5}{7}c_1 = -\frac{6}{7}c_0$$

$$c_3 = -\frac{7}{\frac{27}{2}}c_2 = -\frac{4}{9}c_0$$

etc. and the first Frobenius series is

$$u_1 = c_0\left(x - \frac{6}{5}x^2 + \frac{6}{7}x^3 - \frac{4}{9}x^4 + \cdots\right) \qquad (4.30)$$

Setting $r = -1/2$ in the recurrence equation (4.26) and using b_n instead of c_n to distinguish it from the first case,

$$b_n = -\frac{2n-2}{n\left(n - \frac{3}{2}\right)}b_{n-1} \qquad (4.31)$$

Noting that in this case $b_1 = 0$, all the following b_ns must be zero and the second Frobenius series has only one term: $b_0 x^{-1/2}$. The complete solution is

$$u = c_0\left(x - \frac{6}{5}x^2 + \frac{6}{7}x^3 - \frac{4}{9}x^4 + \cdots\right) + b_0 x^{-1/2} \qquad (4.32)$$

Example 4.4 (repeated roots). Next consider the differential equation

$$x^2 u'' - xu' + (x+1)u = 0 \qquad (4.33)$$

There is a regular singular point at $x = 0$, so we attempt a Frobenius series around $x = 0$.

Differentiating (4.17) and substituting into (4.30),

$$\sum_{n=0}^{\infty}(n+r-1)(n+r)c_n x^{n+r} - \sum_{n=0}^{\infty}(n+r)c_n x^{n+r} + \sum_{n=0}^{\infty}c_n x^{n+r} + \sum_{n=0}^{\infty}c_n x^{n+r+1} = 0 \quad (4.34)$$

or

$$[r(r-1) - r + 1]c_0 x^r + \sum_{n=1}^{\infty}[(n+r-1)(n+r) - (n+r) + 1]c_n x^{n+r} + \sum_{n=1}^{\infty}c_{n-1}x^{n+r} = 0$$

$$(4.35)$$

where we have shifted the index in the last sum.

The indicial equation is

$$r(r-1) - r + 1 = 0 \qquad (4.36)$$

and the roots of this equation are both $r = 1$. Setting the last two sums to zero we find the recurrence equation

$$c_n = -\frac{1}{(n+r-1)(n+r) - (n+r) + 1} c_{n-1} \tag{4.37}$$

and since $r = 1$,

$$c_n = -\frac{1}{n(n+1) - (n+1) + 1} c_{n-1} \tag{4.38}$$

$$c_1 = -c_0$$

$$c_2 = \frac{-1}{6 - 3 + 1} c_1 = \frac{1}{4} c_0$$

$$c_3 = \frac{-1}{12 - 4 + 1} c_2 = \frac{-1}{9} c_1 = \frac{-1}{36} c_0$$

etc.

The Frobenius series is

$$u_1 = c_0 \left(x - x^2 + \frac{1}{4} x^3 - \frac{1}{36} x^4 + \cdots \right) \tag{4.39}$$

In this case there is no second solution in the form of a Frobenius series because of the repeated root. We shall soon see what form the second solution takes.

Example 4.5 (roots differing by an integer 1). Next consider the equation

$$x^2 u'' - 2xu' + (x+2)u = 0 \tag{4.40}$$

There is a regular singular point at $x = 0$. We therefore expect a solution in the form of the Frobenius series (4.18). Substituting (4.18), (4.19), (4.20) into our differential equation, we obtain

$$\sum_{n=0}^{\infty}(n+r)(n+r-1)c_n x^{n+r} - \sum_{n=0}^{\infty} 2(n+r)c_n x^{n+r} + \sum_{n=0}^{\infty} 2c_n x^{n+r} + \sum_{n=0}^{\infty} c_n x^{n+r+1} = 0 \tag{4.41}$$

Taking out the $n = 0$ term and shifting the last summation,

$$[r(r-1) - 2r + 2]c_0 x^r + \sum_{n=1}^{\infty}[(n+r)(n+r-1) - 2(n+r) + 2]c_n x^{n+r}$$

$$+ \sum_{n=1}^{\infty} c_{n-1} x^{n+r} = 0 \tag{4.42}$$

The first term is the indicial equation.

$$r(r-1) - 2r + 2 = 0 \qquad (4.43)$$

There are two distinct roots, $r_1 = 2$ and $r_2 = 1$. However they differ by an integer.

$$r_1 - r_2 = 1.$$

Substituting $r_1 = 2$ into (4.39) and noting that each coefficient of x^{n+r} must be zero,

$$[(n+2)(n+1) - 2(n+2) + 2]c_n + c_{n-1} = 0 \qquad (4.44)$$

The recurrence equation is

$$c_n = \frac{-c_{n-1}}{(n+2)(n-1)+2}$$

$$c_1 = \frac{-c_0}{2}$$

$$c_2 = \frac{-c_1}{6} = c_0 \frac{c_0}{12}$$

$$c_3 = \frac{-c_2}{12} = \frac{-c_0}{144} \qquad (4.45)$$

The first Frobenius series is therefore

$$u_1 = c_0 \left[x^2 - \frac{1}{2}x^3 + \frac{1}{12}x^4 - \frac{1}{144}x^5 + \dots \right] \qquad (4.46)$$

We now attempt to find the Frobenius series corresponding to $r_2 = 1$. Substituting into (4.44) we find that

$$[n(n+1) - 2(n+1) + 2]c_n = -c_{n-1} \qquad (4.47)$$

When $n = 1$, c_0 must be zero. Hence c_n must be zero for all n and the attempt to find a second Frobenius series has failed. This will not always be the case when roots differ by an integer as illustrated in the following example.

Example 4.6 (roots differing by an integer 2).   Consider the differential equation

$$x^2 u'' + x^2 u' - 2u = 0 \qquad (4.48)$$

You may show that the indicial equation is $r^2 - r - 2 = 0$ with roots $r_1 = 2, r_2 = -1$ and the roots differ by an integer. When $r = 2$ the recurrence equation is

$$c_n = -\frac{n+1}{n(n+3)}c_{n-1} \qquad (4.49)$$

The first Frobenius series is

$$u_1 = c_0 x^2 \left[1 - \frac{1}{2}x + \frac{3}{20}x^2 - \frac{1}{30}x^3 + \cdots \right] \tag{4.50}$$

When $r = -1$ the recurrence equation is

$$[(n-1)(n-2) - 2]b_n + (n-2)b_{n-1} = 0 \tag{4.51}$$

When $n = 3$ this results in $b_2 = 0$. Thus $b_n = 0$ for all $n \geq 2$ and the second series terminates.

$$u_2 = b_0 \left(\frac{1}{x} - \frac{1}{2} \right) \tag{4.52}$$

4.1.5   Lessons: Finding Series Solution for Differential Equations with Regular Singular Points

1. Assume a solution of the form

$$u = \sum_{n=0}^{\infty} c_n x^{n+r}, \, c_0 \neq 0 \tag{4.53}$$

Differentiate term by term and insert into the differential equation. Set the coefficient of the lowest power of x to zero to obtain a quadratic equation on r.

If the indicial equation yields two roots that do not differ by an integer there will always be two Frobenius series, one for each root of the indicial equation.

2. If the roots are the same (repeated roots) the form of the second solution will be

$$u_2 = u_1 \ln(x) + \sum_{n=1}^{\infty} b_n x^{n+r_1} \tag{4.54}$$

This equation is substituted into the differential equation to determine b_n.

3. If the roots differ by an integer, choose the largest root to obtain a Frobenius series for u_1. The second solution may be another Frobenius series. If the method fails assume a solution of the form

$$u_2 = u_1 \ln(x) + \sum_{n=1}^{\infty} b_n x^{n+r_2} \tag{4.55}$$

This equation is substituted into the differential equation to find b_n.

This is considered in the next section.

4.1.6   Logarithms and Second Solutions

Example 4.7.   Reconsider Example 4.4 and assume a solution in the form of (4.54). Recall that in Example 4.4 the differential equation was

$$x^2 u'' - xu' + (1+x)u = 0 \qquad (4.56)$$

and the indicial equation yielded a double root at $r = 1$.

A single Frobenius series was

$$u_1 = x - x^2 + \frac{x^3}{4} - \frac{x^4}{36} + \cdots$$

Now differentiate Eq. (4.54).

$$u_2' = u_1' \ln x + \frac{1}{x}u_1 + \sum_{n=1}^{\infty}(n+r)b_n x^{n+r-1}$$

$$u_2'' = u_1'' \ln x + \frac{2}{x}u_1' - \frac{1}{x^2}u_1 + \sum_{n=1}^{\infty}(n+r-1)(n+r)b_n x^{n+r-2} \qquad (4.57)$$

Inserting this into the differential equation gives

$$\ln(x)[x^2 u_1'' - xu_1' + (1+x)u_1] + 2(xu_1' - u_1)$$

$$+ \sum_{n=1}^{\infty}[b_n(n+r-1)(n+r)x^{n+r} - b_n(n+r)x^{n+r} + b_n x^{n+r}]$$

$$+ \sum_{n=1}^{\infty} b_n x^{n+r+1} = 0 \qquad (4.58)$$

The first term on the left-hand side of (4.52) is clearly zero because the term in brackets is the original equation. Noting that $r = 1$ in this case and substituting from the Frobenius series for u_1, we find (c_0 can be set equal to unity without losing generality)

$$2\left[-x^2 + \frac{x^3}{3} - \frac{x^4}{12} + \cdots\right] + \sum_{n=1}^{\infty}[n(n+1) - (n+1) + 1]b_n x^{n+1} + \sum_{n=2}^{\infty} b_{n-1}x^{n+1} = 0 \qquad (4.59)$$

or

$$-2x^2 + x^3 - \frac{x^4}{6} + \cdots + b_1 x^2 + \sum_{n=2}^{\infty}[n^2 b_n + b_{n-1}]x^{n+1} = 0 \qquad (4.60)$$

Equating coefficients of x raised to powers we find that $b_1 = 2$

For $n \geq 2$

$$1 + 4b_2 + b_1 = 0 \qquad b_2 = -3/4$$

$$-\frac{1}{6} + 9b_3 + b_2 = 0 \qquad b_3 = \frac{11}{108}$$

etc.

$$u_2 = u_1 \ln x + \left(2x^2 - \frac{3}{4}x^3 + \frac{11}{108}x^4 - \cdots \right) \qquad (4.61)$$

The complete solution is

$$u = [C_1 + C_2 \ln x] u_1 + C_2 \left[2x^2 - \frac{3}{4}x^3 + \frac{11}{108}x^4 - \cdots \right] \qquad (4.62)$$

Example 4.8.   Reconsider Example 4.5 in which a second Frobenius series could not be found because the roots of the indicial equation differed by an integer. We attempt a second solution in the form of (4.55).

The differential equation in Example 4.5 was

$$x^2 u'' - 2xu' + (x + 2)u = 0$$

and the roots of the indicial equation were $r = 2$ and $r = 1$, and are therefore separated by an integer. We found one Frobenius series

$$u_1 = x^2 - \frac{1}{2}x^3 + \frac{1}{12}x^4 - \frac{1}{144}x^5 + \cdots$$

for the root $r = 2$, but were unable to find another Frobenius series for the case of $r = 1$.

Assume a second solution of the form in Eq. (4.55). Differentiating and substituting into (4.40)

$$[x^2 u_1'' - 2xu' + (x + 2)u] \ln(x) + 2xu' - 3u_1$$

$$+ \sum_{n=1}^{\infty} b_n [(n + r)(n + r - 1) - 2(n + r) + 2]x^{n+r}$$

$$+ \sum_{n=1}^{\infty} b_n x^{n+r+1} = 0 \qquad (4.63)$$

Noting that the first term in the brackets is zero, inserting u_1 and u_1' from (4.50) and noting that $r_2 = 1$

$$x^2 - \frac{3}{2}x^3 + \frac{5}{12}x^4 - \frac{7}{144}x^5 + \ldots + b_0 x^2 + \sum_{n=2}^{\infty} \{[n(n-1)]b_n + b_{n-1}\}x^{n+1} = 0 \qquad (4.64)$$

Equating x^2 terms, we find that $b_0 = -1$. For higher order terms

$$\frac{3}{2} = 2b_2 + b_1 = 2b_2 + b_1$$

Taking $b_1 = 0$,

$$b_2 = \frac{3}{4}$$

$$-\frac{5}{12} = 6b_3 + b_2 = 6b_3 + \frac{3}{4}$$

$$b_3 = -\frac{7}{36}$$

The second solution is

$$u_2 = u_1 \ln(x) - \left(x - \frac{3}{4}x^3 + \frac{7}{36}x^4 - \cdots \right)$$ (4.65)

The complete solution is therefore

$$u = [C_1 + C_2 \ln x] u_1 - C_2 \left[x - \frac{3}{4}x^3 + \frac{7}{36}x^4 - \cdots \right]$$ (4.66)

Problems

1. Find two Frobenius series solutions

$$x^2 u'' + 2xu' + (x^2 - 2)u = 0$$

2. Find two Frobenious series solutions

$$x^2 u'' + xu' + \left(x^2 - \frac{1}{4} \right) u = 0$$

3. Show that the indicial equation for the differential equation

$$xu'' + u' + xu = 0$$

has roots $s = -1$ and that the differential equation has only one Frobenius series solution. Find that solution. Then find another solution in the form

$$u = \ln \sum_{n=0}^{\infty} c_n x^{n+s} + \sum_{m=0}^{\infty} a_n x^{s+m}$$

where the first summation above is the first Frobenius solution.

4.2   BESSEL FUNCTIONS

A few differential equations are so widely useful in applied mathematics that they have been named after the mathematician who first explored their theory. Such is the case with Bessel's equation. It occurs in problems involving the Laplacian $\nabla^2 u$ in cylindrical coordinates when variables are separated. Bessel's equation is a Sturm–Liouville equation of the form

$$\rho^2 \frac{d^2 u}{d\rho^2} + \rho \frac{du}{d\rho} + (\lambda^2 \rho^2 - v^2)u = 0 \qquad (4.67)$$

Changing the independent variable $x = \lambda\rho$, the equation becomes

$$x^2 u'' + x u' + (x^2 - v^2)u = 0 \qquad (4.68)$$

4.2.1   Solutions of Bessel's Equation

Recalling the standard forms (4.1) and (4.2) we see that it is a linear homogeneous equation with variable coefficients and with a regular singular point at $x = 0$. We therefore assume a solution of the form of a Frobenius series (4.17).

$$u = \sum_{j=0}^{\infty} c_j x^{j+r} \qquad (4.69)$$

Upon differentiating twice and substituting into (4.68) we find

$$\sum_{j=0}^{\infty}[(j+r-1)(j+r)+(j+r)-v^2]c_j x^{j+r} + \sum_{j=0}^{\infty} c_j x^{j+r+2} = 0 \qquad (4.70)$$

In general v can be any real number. We will first explore some of the properties of the solution when v is a nonnegative integer, $0, 1, 2, 3, \ldots$. First note that

$$(j+r-1)(j+r)+(j+r) = (j+r)^2 \qquad (4.71)$$

Shifting the exponent in the second summation and writing out the first two terms in the first

$$(r-n)(r+n)c_0 + (r+1-n)(r+1+n)c_1 x$$

$$+ \sum_{j=2}^{\infty}[(r+j-n)(r+j+n)c_j + c_{j-2}]x^j = 0 \qquad (4.72)$$

In order for the coefficient of the x^0 term to vanish $r = n$ or $r = -n$. (This is the indicial equation.) In order for the coefficient of the x term to vanish $c_1 = 0$. For each term in the summation to vanish

$$c_j = \frac{-1}{(r+j-n)(r+j+n)}c_{j-2} = \frac{-1}{j(2n+j)}c_{j-2}, \qquad r=n \qquad j=2,3,4,\cdots \quad (4.73)$$

This is the recurrence relation. Since $c_1 = 0$, all $c_j = 0$ when j is an odd number. It is therefore convenient to write $j = 2k$ and note that

$$c_{2k} = \frac{-1}{2^2 k(r+k)}c_{2k-2} \qquad (4.74)$$

so that

$$c_{2k} = \frac{(-1)^k}{k!(n+1)(n+2)\ldots(n+k)2^{2k}}c_0 \qquad (4.75)$$

The Frobenius series is

$$u = c_0 x^n \left[1 + \sum_{k=1}^{\infty} \frac{(-1)^k}{k!(n+1)(n+2)\ldots.(n+k)}\left(\frac{x}{2}\right)^{2k}\right] \qquad (4.76)$$

Now c_0 is an arbitrary constant so we can choose it to be $c_0 = 1/n!2^n$ in which case the above equation reduces to

$$J_n = u = \sum_{k=0}^{\infty} \frac{(-1)^k}{k!(n+k)!}\left(\frac{x}{2}\right)^{n+2k} \qquad (4.77)$$

The usual notation is J_n and the function is called a *Bessel function of the first kind of order n*. Note that we can immediately conclude from (4.77) that

$$J_n(-x) = (-1)^n J_n(x) \qquad (4.78)$$

Note that the roots of the indicial equation differ by an integer. When $r = -n$ (4.72) does not yield a useful second solution since the denominator is zero for $j = 0$ or $2n$. In any case it is easy to show that $J_n(x) = (-1)^n J_{-n}$, so when r is an integer the two solutions are not independent.

A second solution is determined by the methods detailed above and involves natural logarithms. The details are very messy and will not be given here. The result is

$$Y_n(x) = \frac{2}{\pi}\left\{J_n(x)\left[\ln\left(\frac{x}{2}\right) + \gamma\right] + \sum_{k=1}^{\infty}\frac{(-1)^{k+1}[\phi(k) + \phi(k+1)]}{2^{2k+n+1}k!(k+n)!}x^{2k+n}\right\}$$

$$-\frac{2}{\pi}\sum_{k=0}^{n-1}\frac{(n-k-1)!}{2^{2k-n+1}k!}x^{2k-n} \qquad (4.79)$$

In this equation $\Phi(k) = 1 + 1/2 + 1/3 + \cdots + 1/k$ and γ is Euler's constant 0.5772156649 ......

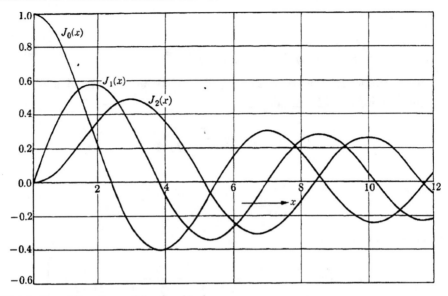

FIGURE 4.1: Bessel functions of the first kind

Bessel functions of the first and second kinds of order zero are particularly useful in solving practical problems (Fig. 4.1). For these cases

$$J_0(x) = \sum_{k=0}^{\infty} \frac{(-1)^k}{(k!)^2} \left(\frac{x}{2}\right)^{2k} \qquad (4.80)$$

and

$$Y_0 = J_0(x)\ln(x) + \sum_{k=1}^{\infty} \frac{(-1)^{k+1}}{2^{2k}(k!)^2}\phi(k)x^{2k} \qquad (4.81)$$

The case of $v \neq n$.   Recall that in (4.70) if v is not an integer, a part of the denominator is

$$(1 + v)(2 + v)(3 + v)\ldots(n + v) \qquad (4.82)$$

We were then able to use the familiar properties of *factorials* to simplify the expression for $J_n(x)$. If $v \neq n$ we can use the properties of the *gamma function* to the same end. The gamma function is defined as

$$\Gamma(v) = \int_0^{\infty} t^{v-1}e^{-t}dt \qquad (4.83)$$

Note that

$$\Gamma(v+1) = \int_0^\infty t^v e^{-t} dt \tag{4.84}$$

and integrating by parts

$$\Gamma(v+1) = [-t\,v e^{-t}]_0^\infty + v \int_0^\infty t^{v-1} e^{-t} dt = v\Gamma(v) \tag{4.85}$$

and (4.82) can be written as

$$(1+v)(2+v)(3+v)\ldots.(n+v) = \frac{\Gamma(n+v+1)}{\Gamma(v+1)} \tag{4.86}$$

so that when v is not an integer

$$J_v(x) = \sum_{n=0}^\infty \frac{(-1)^n}{2^{2n+v} n! \Gamma(n+v+1)} x^{2n+v} \tag{4.87}$$

Fig. 4.3 is a graphical representation of the gamma function.

Here are the rules

1. If $2v$ is not an integer, J_v and J_{-v} are linearly independent and the general solution of Bessel's equation of order v is

$$u(x) = AJ_v(x) + BJ_{-v}(x) \tag{4.88}$$

    where A and B are constants to be determined by boundary conditions.

2. If $2v$ is an odd positive integer J_v and J_{-v} are still linearly independent and the solution form (4.88) is still valid.

3. If $2v$ is an even integer, $J_v(x)$ and $J_{-v}(x)$ are not linearly independent and the solution takes the form

$$u(x) = AJ_v(x) + BY_v(x) \tag{4.89}$$

Bessel functions are tabulated functions, just as are exponentials and trigonometric functions. Some examples of their shapes are shown in Figs. 4.1 and 4.2.

Note that both $J_v(x)$ and $Y_v(x)$ have an infinite number of zeros and we denote them as $\lambda_j, j = 0, 1, 2, 3, \ldots$

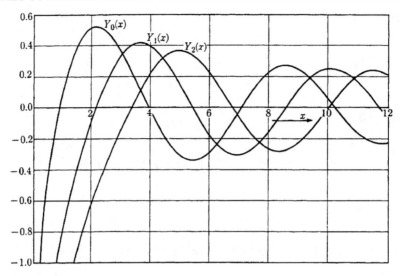

FIGURE 4.2: Bessel functions of the second kind

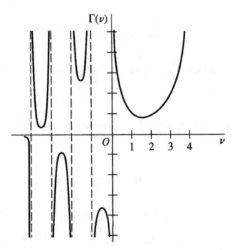

FIGURE 4.3: The gamma function

Some important relations involving Bessel functions are shown in Table 4.1. We will derive only the first, namely

$$\frac{d}{dx}(x^{v}J_{v}(x)) = x^{v}J_{v-1}(x) \tag{4.90}$$

$$\frac{d}{dx}(x^{v}J_{v}(x)) = \frac{d}{dx}\left[\sum_{n=0}^{\infty}\frac{(-1)^{n}}{2^{2n+v}n!\Gamma(n+v+1)}x^{2n+2v}\right] \tag{4.91}$$

TABLE 4.1: Some Properties of Bessel Functions

1. $[x^\nu J_\nu(x)]' = x^\nu J_{\nu-1}(x)$
2. $[\vec{x}^{-\nu} J_\nu(x)]' = -x^{-\nu} J_{\nu+1}(x)$
3. $J_{\nu-1}(x) + J_{\nu+1}(x) = 2\nu/x[J_\nu(x)]$
4. $J_{\nu-1}(x) - J_{\nu+1}(x) = 2J_\nu(x)'$
5. $\int x^\nu J_{\nu-1}(x)dx = x^\nu J_\nu + constant$
6. $\int x^{-\nu} J_{\nu+1}(x)dx = x^{-\nu} J_\nu(x) + constant$

$$= \sum_{n=0}^{\infty} \frac{(-1)^n 2(n+\nu)}{2^{2n+\nu} n!(n+\nu)\Gamma(n+\nu)} x^{2n+2\nu-1} \qquad (4.92)$$

$$= x^\nu \sum_{n=0}^{\infty} \frac{(-1)^n}{2^{2n+\nu-1} n!\Gamma(n+\nu)} x^{2n+2\nu-1} = x^\nu J_{\nu-1}(x) \qquad (4.93)$$

These will prove important when we begin solving partial differential equations in cylindrical coordinates using separation of variables.

Bessel's equation is of the form (4.138) of a Sturm–Liouville equation and the functions $J_n(x)$ are orthogonal with respect to a weight function ρ (see Eqs. (3.46) and (3.53), Chapter 3).

Note that Bessel's equation (4.67) with $\nu = n$ is

$$\rho^2 J_n'' + \rho J_n' + (\lambda^2 \rho^2 - n^2)J_n = 0 \qquad (4.94)$$

which can be written as

$$\frac{d}{d\rho}(\rho J_n')^2 + (\lambda^2 \rho^2 - n^2)\frac{d}{d\rho}J_n^2 = 0 \qquad (4.95)$$

Integrating, we find that

$$[(\rho J')^2 + (\lambda^2 \rho^2 - n^2)J^2]_0^1 - 2\lambda^2 \int_{\rho=0}^{1} \rho J^2 d\rho = 0 \qquad (4.96)$$

Thus,

$$2\lambda^2 \int_{\rho=0}^{1} \rho J_n^2 d\rho = \lambda^2 [J_n'(\lambda)]^2 + (\lambda^2 - n^2)[J_n(\lambda)]^2 \qquad (4.97)$$

Thus, we note from that if the eigenvalues are λ_j, the roots of $J_\nu(\lambda_j\rho) = 0$ the orthogonality condition is, according to Eq. (3.53) in Chapter 3

$$\int_0^1 \rho J_n(\lambda_j\rho)J_n(\lambda_k\rho)d\rho = 0, \qquad j \neq k$$

$$= \frac{1}{2}[J_{n+1}(\lambda_j)]^2, \quad j = k \qquad (4.98)$$

On the other hand, if the eigenvalues are the roots of the equation

$$HJ_n(\lambda_j) + \lambda_j J_n'(\lambda_j) = 0$$

$$\int_0^1 \rho J_n(\lambda_j\rho)J_n(\lambda_k\rho)d\rho = 0, \quad j \neq k$$

$$= \frac{(\lambda_j^2 - n^2 + H^2)[J_n(\lambda_j)]^2}{2\lambda_j^2}, \quad j = k \qquad (4.99)$$

Using the equations in the table above and integrating by parts it is not difficult to show that

$$\int_{s=0}^x s^n J_0(s)ds = x^n J_1(x) + (n-1)x^{n-1}J_0(x) - (n-1)^2 \int_{s=0}^x s^{n-2}J_0(s)ds \qquad (4.100)$$

4.2.2   Fourier–Bessel Series

Owing to the fact that Bessel's equation with appropriate boundary conditions is a Sturm–Liouville system it is possible to use the orthogonality property to expand any piecewise continuous function on the interval $0 < x < 1$ as a series of Bessel functions. For example, let

$$f(x) = \sum_{n=1}^\infty A_n J_0(\lambda_n x) \qquad (4.101)$$

Multiplying both sides by $xJ_0(\lambda_k x)dx$ and integrating from $x = 0$ to $x = 1$ (recall that the weighting function x must be used to insure orthogonality) and noting the orthogonality property we find that

$$f(x) = \sum_{j=1}^\infty \frac{\int_{x=0}^1 xf(x)J_0(\lambda_j x)dx}{\int_{x=0}^1 x[J_0(\lambda_j x)]^2 dx} J_0(\lambda_j x) \qquad (4.102)$$

Example 4.9. Derive a Fourier–Bessel series representation of 1 on the interval $0 < x < 1$. We note that with $J_0(\lambda_j) = 0$

$$\int_{x=0}^{1} x[J_0(\lambda_j x)]^2 dx = \frac{1}{2}[J_1(\lambda_j)]^2 \qquad (4.103)$$

and

$$\int_{x=0}^{1} x J_0(\lambda_j x) dx = J_1(\lambda_j) \qquad (4.104)$$

Thus

$$1 = 2 \sum_{j=1}^{\infty} \frac{J_0(\lambda_j x)}{\lambda_j J_1(\lambda_j)} \qquad (4.105)$$

Example 4.10 (A problem in cylindrical coordinates). A cylinder of radius r_1 is initially at a temperature u_0 when its surface temperature is increased to u_1. It is sufficiently long that variation in the z direction may be neglected and there is no variation in the θ direction. There is no heat generation. From Chapter 1, Eq. (1.11)

$$u_t = \frac{\alpha}{r}(r u_r)_r \qquad (4.106)$$

$$u(0, r) = u_0$$

$$u(t, r_1) = u_1$$

$$u \text{ is bounded} \qquad (4.107)$$

The length scale is r_1 and the time scale is r_1^2/α. A dimensionless dependent variable that normalizes the problem is $(u - u_1)/(u_0 - u_1) = U$. Setting $\eta = r/r_1$ and $\tau = t\alpha/r_1^2$,

$$U_\tau = \frac{1}{\eta}(\eta U_\eta)_\eta \qquad (4.108)$$

$$U(0, \eta) = 1$$

$$U(\tau, 1) = 0 \qquad (4.109)$$

$$U \text{ is bounded}$$

Separate variables as $T(\tau)R(\eta)$. Substitute into the differential equation and divide by TR.

$$\frac{T_\tau}{T} = \frac{1}{R\eta}(\eta R_\eta)_\eta = \pm\lambda^2 \qquad (4.110)$$

where the minus sign is chosen so that the function is bounded. The solution for T is exponential and we recognize the equation for R as Bessel's equation with $\nu = 0$.

$$\frac{1}{\eta}(\eta R_\eta)_\eta + \lambda^2 R = 0 \qquad (4.111)$$

The solution is a linear combination of the two Bessel functions of order 0.

$$C_1 J_0(\lambda\eta) + C_2 Y_0(\lambda\eta) \qquad (4.112)$$

Since we have seen that Y_0 is unbounded as η approaches zero, C_2 must be zero. Furthermore, the boundary condition at $\eta = 1$ requires that $J_0(\lambda) = 0$, so that our eigenfunctions are $J_0(\lambda\eta)$ and the corresponding eigenvalues are the roots of $J_0(\lambda_n) = 0$.

$$U_n = K_n e^{-\lambda_n^2 \tau} J_0(\lambda_n\eta), \qquad n = 1, 2, 3, 4, \ldots \qquad (4.113)$$

Summing (linear superposition)

$$U = \sum_{n=1}^{\infty} K_n e^{-\lambda_n^2 \tau} J_0(\lambda_n\eta) \qquad (4.114)$$

Using the initial condition,

$$1 = \sum_{n=1}^{\infty} K_n J_0(\lambda_n\eta) \qquad (4.115)$$

Bessel functions are orthogonal *with respect to weighting factor* η since they are solutions to a Sturm–Liouville system. Therefore when we multiply both sides of this equation by $\eta J_0(\lambda_m\eta)d\eta$ and integrate over $(0, 1)$ all of the terms in the summation are zero except when $m = n$. Thus,

$$\int_{\eta=0}^{1} J_0(\lambda_n\eta)\eta d\eta = K_n \int_{\eta=0}^{1} J_0^2(\lambda_n\eta)\eta d\eta \qquad (4.116)$$

but

$$\int_{\eta=0}^{1} \eta J_0^2(\lambda_n\eta)d\eta = \frac{J_1^2(\lambda_n)}{2}$$

$$\int_{\eta=0}^{1} \eta J_0(\lambda_n\eta)d\eta = \frac{1}{\lambda_n}J_1(\lambda_n) \qquad (4.117)$$

Thus

$$U(\tau, \eta) = \sum_{n=0}^{\infty} \frac{2}{\lambda_n J_1(\lambda_n)} e^{-\lambda_n^2 \tau} J_0(\lambda_n \eta) \qquad (4.118)$$

Example 4.11 (Heat generation in a cylinder). Reconsider the problem of heat transfer in a long cylinder but with heat generation and at a normalized initial temperature of zero.

$$u_\tau = \frac{1}{r}(r u_r)_r + q_0 \qquad (4.119)$$

$$u(\tau, 1) = u(0, r) = 0, \ u \text{ bounded} \qquad (4.120)$$

Our experience with the above example hints that the solution maybe of the form

$$u = \sum_{j=1}^{\infty} A_j(\tau) J_0(\lambda_j r) \qquad (4.121)$$

This equation satisfies the boundary condition at $r = 1$ and $A_j(\tau)$ is to be determined. Substituting into the partial differential equation gives

$$\sum_{j=1}^{\infty} A_j'(\tau) J_0(\lambda_j) = \sum_{j=1}^{\infty} A_j(\tau) \frac{1}{r} \frac{d}{dr} \left[r \frac{dJ_0}{dr} \right] + q_0 \qquad (4.122)$$

In view of Bessel's differential equation, the first term on the right can be written as

$$\sum_{j=1}^{\infty} -\lambda_j^2 J_0(\lambda_j r) A_j(\tau) \qquad (4.123)$$

The second term can be represented as a Fourier–Bessel series as follows:

$$q_0 = q_0 \sum_{j=1}^{\infty} \frac{2 J_0(\lambda_j r)}{\lambda_j J_1(\lambda_j)} \qquad (4.124)$$

as shown in Example 4.9 above.

Equating coefficients of $J_0(\lambda_j r)$ we find that $A_j(\tau)$ must satisfy the ordinary differential equation

$$A'(\tau) + \lambda_j^2 A(\tau) = q_0 \frac{2}{\lambda_j J_1(\lambda_j)} \qquad (4.125)$$

with the initial condition $A(0) = 0$.

Solution of this simple first-order linear differential equations yields

$$A_j(\tau) = \frac{2 q_0}{\lambda_j^3 J_1(\lambda_j)} + C \exp(-\lambda_j^2 \tau) \qquad (4.126)$$

After applying the initial condition

$$A_j(\tau) = \frac{2q_0}{\lambda_j^3 J_1(\lambda_j)} \left[1 - \exp(-\lambda_j^2 \tau)\right] \tag{4.127}$$

The solution is therefore

$$u(\tau, r) = \sum_{j=1}^{\infty} \frac{2q_0}{\lambda_j^3 J_1(\lambda_j)} \left[1 - \exp(-\lambda_j^2 \tau)\right] J_0(\lambda_j r) \tag{4.128}$$

Example 4.12 (Time dependent heat generation). Suppose that instead of constant heat generation, the generation is time dependent, $q(\tau)$. The differential equation for $A(\tau)$ then becomes

$$A'(\tau) + \lambda_j^2 A(\tau) = \frac{2q(\tau)}{\lambda_j J_1(\lambda_j)} \tag{4.129}$$

An integrating factor for this equation is $\exp(\lambda_j^2 \tau)$ so that the equation can be written as

$$\frac{d}{d\tau}\left[A_j \exp(\lambda_j^2 \tau)\right] = \frac{2q(\tau)}{\lambda_j J_1(\lambda_j)} \exp(\lambda_j^2 \tau) \tag{4.130}$$

Integrating and introducing as a dummy variable t

$$A_j(\tau) = \frac{2}{\lambda_j J_1(\lambda_j)} \int_{t=0}^{\tau} q(t) \exp(-\lambda_j^2(\tau - t)) dt \tag{4.131}$$

Problems

1. By differentiating the series form of $J_0(x)$ term by term show that

$$J_0'(x) = -J_1(x)$$

2. Show that

$$\int x J_0(x) dx = x J_1(x) + constant$$

3. Using the expression for $\int_{s=0}^{x} s^n J_0(s) ds$ show that

$$\int_{s=0}^{x} s^5 J_0(s) ds = x(x^2 - 8)[4x J_0(x) + (x^2 - 8) J_1(x)]$$

4. Express $1 - x$ as a Fourier–Bessel series.

4.3   LEGENDRE FUNCTIONS

We now consider another second-order linear differential that is common for problems involving the Laplacian in spherical coordinates. It is called Legendre's equation,

$$(1 - x^2)u'' - 2xu' + ku = 0 \qquad (4.132)$$

This is clearly a Sturm–Liouville equation and we will seek a series solution near the origin, which is a regular point. We therefore assume a solution in the form of (4.3).

$$u = \sum_{j=0}^{\infty} c_j x^j \qquad (4.133)$$

Differentiating (4.133) and substituting into (4.132) we find

$$\sum_{j=0}^{\infty} [j(j-1)c_j x^{j-2}(1-x^2) - 2jc_j x^j + n(n+1)c_j x^j] \qquad (4.134)$$

or

$$\sum_{j=0}^{\infty} \{[k - j(j+1)]c_j x^j + j(j-1)c_j x^{j-2}\} = 0 \qquad (4.135)$$

On shifting the last term,

$$\sum_{j=0}^{\infty} \{(j+2)(j+1)c_{j+2} + [k - j(j+1)]c_j\}x^j = 0 \qquad (4.136)$$

The recurrence relation is

$$c_{j+2} = -\frac{j(j+1) - k}{(j+1)(j+2)}c_j \qquad (4.137)$$

There are thus two independent Frobenius series. It can be shown that they both diverge at $x = 1$ unless they terminate at some point. It is easy to see from (4.137) that they do in fact terminate if $k = n(n+1)$.

Since n and j are integers it follows that $c_{n+2} = 0$ and consequently c_{n+4}, c_{n+6}, etc. are all zero. Therefore the solutions, which depend on n (i.e., the eigenfunctions) are polynomials, series that terminate at $j = n$. For example, if $n = 0$, $c_2 = 0$ and the solution is a constant. If

$n = 1$ $c_n = 0$ when $n \geq 1$ and the polynomial is x. In general

$$u = P_n(x) = c_n \left[x^n - \frac{n(n-1)}{2(2n-1)} x^{n-2} + \frac{n(n-1)(n-2)(n-3)}{2(4)(2n-1)(2n-3)} x^{n-4} - \cdots \right]$$

$$= \frac{1}{2^k} \sum_{k=0}^{m} \frac{(-1)^k}{k!} \frac{(2n-2k)!}{(n-2k)!(n-k)!} x^{n-2k} \qquad (4.138)$$

where $m = n/2$ if n is even and $(n-1)/2$ if n is odd.

The coefficient c_n is of course arbitrary. It turns out to be convenient to choose it to be

$$c_0 = 1$$
$$c_n = \frac{(2n-1)(2n-3)\cdots 1}{n!} \qquad (4.139)$$

the first few polynomials are

$$P_0 = 1, \ P_1 = x, \ P_2 = (3x^2 - 1)/2, \ P_3 = (5x^3 - 3x)/2, \ P_4 = (35x^4 - 30x^2 + 3)/8,$$

Successive Legendre polynomials can be generated by the use of Rodrigues' formula

$$P_n(x) = \frac{1}{2^n n!} \frac{d^n}{dx^n} (x^2 - 1)^n \qquad (4.140)$$

For example

$$P_5 = (63x^5 - 70x^3 + 15x)/8$$

Fig. 4.4 shows graphs of several Legendre polynomials.

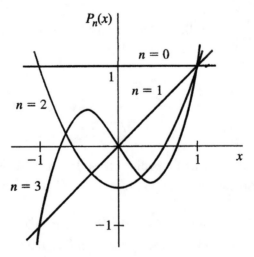

FIGURE 4.4: Legendre polynomials

The second solution of Legendre's equation can be found by the method of variation of parameters. The result is

$$Q_n(x) = P_n(x) \int \frac{d\zeta}{P_n^2(\zeta)(1 - \zeta^2)} \tag{4.141}$$

It can be shown that this generally takes on a logarithmic form involving $\ln[(x + 1)/(x - 1)]$ which goes to infinity at $x = 1$. In fact it can be shown that the first two of these functions are

$$Q_0 = \frac{1}{2} \ln \frac{1 + x}{1 - x} \quad \text{and} \quad Q_1 = \frac{x}{2} \ln \frac{1 + x}{1 - x} - 1 \tag{4.142}$$

Thus the complete solution of the Legendre equation is

$$u = AP_n(x) + BQ_n(x) \tag{4.143}$$

where $P_n(x)$ and $Q_n(x)$ are Legendre polynomials of the first and second kind. If we require the solution to be finite at $x = 1$, B must be zero.

Referring back to Eqs. (3.46) through (3.53) in Chapter 3, we note that the eigenvalues $\lambda = n(n + 1)$ and the eigenfunctions are $P_n(x)$ and $Q_n(x)$. We further note from (3.46) and (3.47) that the weight function is one and that the orthogonality condition is

$$\int_{-1}^{1} P_n(x) P_m(x) dx = \frac{2}{2n + 1} \delta_{mn} \tag{4.144}$$

where δ_{mn} is Kronecker's delta, 1 when $n = m$ and 0 otherwise.

Example 4.13.   Steady heat conduction in a sphere

Consider heat transfer in a solid sphere whose surface temperature is a function of θ, the angle measured downward from the z-axis (see Fig. 1.3 Chapter 1). The problem is steady and there is no heat source.

$$r \frac{\partial^2}{\partial r^2}(ru) + \frac{1}{\sin \theta} \frac{\partial}{\partial \theta} \left(\sin \theta \frac{\partial u}{\partial \theta} \right) = 0$$
$$u(r = 1) = f(\theta) \tag{4.145}$$
$$u \text{ is bounded}$$

Substituting $x = \cos\theta$,

$$r\frac{\partial^2}{\partial r^2}(ru) + \frac{\partial}{\partial x}\left[(1 - x^2)\frac{\partial u}{\partial x}\right] = 0 \qquad (4.146)$$

We separate variables by assuming $u = R(r)X(x)$. Substitute into the equation and divide by RX and find

$$\frac{r}{R}(rR)'' = -\frac{[(1 - x^2)X']'}{X} = \pm\lambda^2 \qquad (4.147)$$

or

$$r(rR)'' \mp \lambda^2 R = 0$$
$$[(1 - x^2)X']' \pm \lambda^2 X = 0 \qquad (4.148)$$

The second of these is Legendre's equation, and we have seen that it has bounded solutions at $r = 1$ when $\lambda^2 = n(n + 1)$. The first equation is of the Cauchy–Euler type with solution

$$R = C_1 r^n + C_2 r^{-n-1} \qquad (4.149)$$

Noting that the constant C_2 must be zero to obtain a bounded solution at $r = 0$, and using superposition,

$$u = \sum_{n=0}^{\infty} K_n r^n P_n(x) \qquad (4.150)$$

and using the condition at $fr = 1$ and the orthogonality of the Legendre polynomial

$$\int_{\theta=0}^{\pi} f(\theta) P_n(\cos\theta)d\theta = \int_{\theta=0}^{\pi} K_n P_n^2(\cos\theta)d\theta = \frac{2K_n}{2n + 1} \qquad (4.151)$$

4.4   ASSOCIATED LEGENDRE FUNCTIONS

Equation (1.15) in Chapter 1 can be put in the form

$$\frac{1}{\alpha}\frac{\partial u}{\partial t} = \left\{\frac{\partial^2 u}{\partial r^2} + \frac{2}{r}\frac{\partial u}{\partial r}\right\} + \frac{1}{r^2}\frac{\partial}{\partial\mu}\left\{(1 - \mu^2)\frac{\partial u}{\partial\mu}\right\} + \frac{1}{r^2(1 - \mu^2)}\frac{\partial^2 u}{\partial\Phi^2} \qquad (4.152)$$

by substituting $\mu = \cos\theta$.

We shall see later that on separating variables in the case where u is a function of r, θ, Φ, and t, we find the following differential equation in the μ variable:

$$\frac{d}{d\mu}\left\{(1-\mu^2)\frac{df}{d\mu}\right\} + \left\{n(n+1) - \frac{m^2}{1-\mu^2}\right\}f = 0 \qquad (4.153)$$

We state without proof that the solution is the associated Legendre function $P_n^m(\mu)$. The associated Legendre polynomial is given by

$$P_n^m = (1-\mu^2)^{1/2m}\frac{d^m}{d\mu^m}P_n(\mu) \qquad (4.154)$$

The orthogonality condition is

$$\int_{-1}^{1}[P_n^m(\mu)]^2 d\mu = \frac{2(n+m)!}{(2n+1)(n-m)!} \qquad (4.155)$$

and

$$\int_{-1}^{1}P_n^m P_{n'}^m d\mu = 0 \qquad n \neq n' \qquad (4.156)$$

The associated Legendre function of the second kind is singular at $x = \pm 1$ and may be computed by the formula

$$Q_n^m(x) = (1-x^2)^{m/2}\frac{d^m Q_n(x)}{dx^m} \qquad (4.157)$$

Problems

1. Find and carefully plot P_6 and P_7.
2. Perform the integral above and show that

$$Q_0(x) = CP_0(x)\int_{\xi=0}^{x}\frac{d\xi}{(1-\xi^2)P_0(\xi)} = \frac{C}{2}\ln\left(\frac{1+x}{1-x}\right)$$

and that

$$Q_1(x) = Cx\int_{\xi=0}^{x}\frac{d\xi}{\xi^2(1-\xi^2)} = \frac{Cx}{2}\ln\left(\frac{1+x}{1-x}\right) - 1$$

3. Using the equation above find $Q_0^0(x)$ and $Q_1^1(x)$

FURTHER READING

J. W. Brown and R. V. Churchill, *Fourier Series and Boundary Value Problems*. New York: McGraw-Hill, 2001.

C. F. Chan Man Fong, D. DeKee, and P. N. Kaloni, *Advanced Mathematics for Engineering and Science*. 2nd edition. Singapore: World Scientific, 2004.

P. V. O'Neil, *Advanced Engineering Mathematics*. 5th edition. Brooks/Cole Thompson, Pacific Grove, CA, 2003.

CHAPTER 5

Solutions Using Fourier Series and Integrals

We have already demonstrated solution of partial differential equations for some simple cases in rectangular Cartesian coordinates in Chapter 2. We now consider some slightly more complicated problems as well as solutions in spherical and cylindrical coordinate systems to further demonstrate the Fourier method of separation of variables.

5.1 CONDUCTION (OR DIFFUSION) PROBLEMS

Example 5.1 (Double Fourier series in conduction). We now consider transient heat conduction in two dimensions. The problem is stated as follows:

$$u_t = \alpha(u_{xx} + u_{yy})$$
$$u(t, 0, y) = u(t, a, y) = u(t, x, 0) = u(t, x, b) = u_0$$
$$u(0, x, y) = f(x, y) \tag{5.1}$$

That is, the sides of a rectangular area with initial temperature $f(x, y)$ are kept at a constant temperature u_0. We first attempt to scale and nondimensionalize the equation and boundary conditions. Note that there are two length scales, a and b. We can choose either, but there will remain an extra parameter, either a/b or b/a in the equation. If we take $\xi = x/a$ and $\eta = y/b$ then (5.1) can be written as

$$\frac{a^2}{\alpha}u_t = \left(u_{\xi\xi} + \frac{a^2}{b^2}u_{\eta\eta}\right) \tag{5.2}$$

The time scale is now chosen as a^2/α and the dimensionless time is $\tau = \alpha t/a^2$. We also choose a new dependent variable $U(\tau, \xi, \eta) = (u - u_0)/(f_{max} - u_0)$. The now nondimensionalized system is

$$U_\tau = U_{\xi\xi} + r^2 U_{\eta\eta} \tag{5.3}$$
$$U(\tau, 0, \eta) = U(\tau, 1, \eta) = U(\tau, \xi, 0) = U(\tau, \xi, 1) = 0$$
$$U(0, \xi, \eta) = (f - u_0)/(f_{max} - u_0) = g(\xi, \eta)$$

We now proceed by separating variables. Let

$$U(\tau, \xi, \eta) = T(\tau)X(\xi)Y(\eta) \tag{5.4}$$

Differentiating and inserting into (5.3) and dividing by (5.4) we find

$$\frac{T'}{T} = \frac{X''Y + r^2 Y'' X}{XY} \tag{5.5}$$

where the primes indicate differentiation with respect to the variable in question and $r = a/b$. Since the left-hand side of (5.5) is a function only of τ and the right-hand side is only a function of ξ and η both sides must be constant. If the solution is to be finite in time we must choose the constant to be negative, $-\lambda^2$. Replacing T'/T by $-\lambda^2$ and rearranging,

$$-\lambda^2 - \frac{X''}{X} = r\frac{Y''}{Y} \tag{5.6}$$

Once again we see that both sides must be constants. How do we choose the signs? It should be clear by now that if either of the constants is positive solutions for X or Y will take the form of hyperbolic functions or exponentials and the boundary conditions on ξ or η cannot be satisfied. Thus,

$$\frac{T'}{T} = -\lambda^2 \tag{5.7}$$

$$\frac{X''}{X} = -\beta^2 \tag{5.8}$$

$$r^2\frac{Y''}{Y} = -\gamma^2 \tag{5.9}$$

Note that X and Y are eigenfunctions of (5.8) and (5.9), which are Sturm–Liouville equations and β and γ are the corresponding eigenvalues.

Solutions of (5.7), (5.8), and (5.9) are

$$T = A\exp(-\lambda^2\tau) \tag{5.10}$$

$$X = B_1 \cos(\beta\xi) + B_2 \sin(\beta\xi) \tag{5.11}$$

$$Y = C_1 \cos(\gamma\eta/r) + C_2 \sin(\gamma\eta/r) \tag{5.12}$$

Applying the first homogeneous boundary condition, we see that $X(0) = 0$, so that $B_1 = 0$. Applying the third homogeneous boundary condition we see that $Y(0) = 0$, so that $C_1 = 0$. The second homogeneous boundary condition requires that $\sin(\beta) = 0$, or $\beta = n\pi$. The last homogeneous boundary condition requires $\sin(\gamma/r) = 0$, or $\gamma = m\pi r$. According to (5.6), $\lambda^2 = \beta^2 + \gamma^2$. Combining these solutions, inserting into (5.4) we have one solution in the form

$$U_{mn}(\tau, \xi, \eta) = K_{nm}e^{-(n^2\pi^2+m^2\pi^2r^2)\tau} \sin(n\pi\xi) \sin(m\pi\eta) \tag{5.13}$$

for all $m, n = 1, 2, 3, 4, 5, \ldots$

Superposition now tells us that

$$\sum_{n=1}^{\infty} \sum_{m=1}^{\infty} K_{nm}e^{-(n^2\pi^2+m^2\pi^2r^2)\tau} \sin(n\pi\xi) \sin(m\pi) \tag{5.14}$$

Using the initial condition

$$g(\xi, \eta) = \sum_{n=1}^{\infty} \sum_{m=1}^{\infty} K_{nm} \sin(n\pi\xi) \sin(m\pi\eta) \tag{5.15}$$

We have a double Fourier series, and since both $\sin(n\pi\xi)$ and $\sin(m\pi\eta)$ are members of orthogonal sequences we can multiply both sides by $\sin(n\pi\xi)\sin(m\pi\eta)d\xi\,d\eta$ and integrate over the domains.

$$\int_{\xi=0}^{1} \int_{\eta=0}^{1} g(\xi, \eta) \sin(n\pi\xi) \sin(m\pi\eta)d\xi\,d\eta$$

$$= K_{nm} \int_{\xi=0}^{1} \int_{\eta=0}^{1} \sin^2(n\pi\xi)d\xi \, \sin^2(m\pi\eta)d\eta$$

$$= \frac{K_{nm}}{4} \tag{5.16}$$

Our solution is

$$\sum_{n=1}^{\infty} \sum_{m=1}^{\infty} 4 \int_{\xi=0}^{1} \int_{\eta=0}^{1} g(\xi, \eta) \sin(n\pi\xi) \sin(m\pi\eta)d\xi\,d\eta \, e^{-(n^2\pi^2+m^2\pi^2r^2)\tau} \sin(n\pi\xi) \sin(m\pi\eta) \tag{5.17}$$

Example 5.2 (A convection boundary condition). Reconsider the problem defined by (2.1) in Chapter 2, but with different boundary and initial conditions,

$$u(t, 0) = u_0 = u(0, x) \tag{5.18}$$

$$ku_x(t, L) - h[u_1 - u(t, L)] = 0 \tag{5.19}$$

The physical problem is a slab with conductivity k initially at a temperature u_0 suddenly exposed at $x = L$ to a fluid at temperature u_1 through a heat transfer coefficient h while the $x = 0$ face is maintained at u_0.

The length and time scales are clearly the same as the problem in Chapter 2. Hence, $\tau = t\alpha/L^2$ and $\xi = x/L$. If we choose $U = (u - u_0)/(u_1 - u_0)$ we make the boundary condition at $x = 0$ homogeneous but the condition at $x = L$ is not. We have the same situation that we had in Section 2.3 of Chapter 2. The differential equation, one boundary condition, and the initial condition are homogeneous. Proceeding, we find

$$U_\tau = U_{\xi\xi}$$
$$U(\tau, 0) = U(0, \xi) = 0 \qquad\qquad (5.20)$$
$$U_\xi(\tau, 1) + B[U(\tau, 1) - 1] = 0$$

where $B = hL/k$. It is useful to relocate the nonhomogeneous condition as the initial condition. As in the previous problem we assume $U(\tau, \xi) = V(\tau, \xi) + W(\xi)$.

$$V_\tau = V_{\xi\xi} + W_{\xi\xi}$$
$$W(0) = 0$$
$$W_\xi(1) + B[W(1) - 1] = 0$$
$$V(\tau, 0) = 0 \qquad\qquad (5.21)$$
$$V_\xi(\tau, 1) + BV(\tau, 1) = 0$$
$$V(0, \xi) = -W(\xi)$$

Set $W_{\xi\xi} = 0$. Integrating twice and using the two boundary conditions on W,

$$W(\xi) = \frac{B\xi}{B + 1} \qquad\qquad (5.22)$$

The initial condition on V becomes

$$V(0, \xi) = -B\xi/(B + 1) . \qquad\qquad (5.23)$$

Assume $V(\tau, \xi) = P(\tau)Q(\xi)$, substitute into the partial differential equation for V, and divide by PQ as usual.

$$\frac{P'}{P} = \frac{Q''}{Q} = \pm\lambda^2 \qquad\qquad (5.24)$$

We must choose the minus sign for the solution to be bounded. Hence,

$$P = Ae^{-\lambda^2\tau}$$
$$Q = C_1 \sin(\lambda\xi) + C_2 \cos(\lambda\xi) \qquad\qquad (5.25)$$

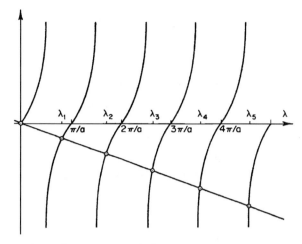

FIGURE 5.1: The eigenvalues of $\lambda_n = -B\tan(\lambda_n)$

Applying the boundary condition at $\xi = 0$, we find that $C_2 = 0$. Now applying the boundary condition on V at $\xi = 1$,

$$C_1\lambda\cos(\lambda) + C_1 B\sin(\lambda) = 0 \tag{5.26}$$

or

$$\lambda = -B\tan(\lambda) \tag{5.27}$$

This is the equation for determining the eigenvalues, λ_n. It is shown graphically in Fig. 5.1.

Example 5.3 (Superposition of several problems). We've seen now that in order to apply separation of variables the partial differential equation itself must be homogeneous and we have also seen a technique for transferring the inhomogeneity to one of the boundary conditions or to the initial condition. But what if several of the boundary conditions are nonhomogeneous? We demonstrate the technique with the following problem. We have a transient two-dimensional problem with given conditions on all four faces.

$$
\begin{aligned}
u_t &= u_{xx} + u_{yy} \\
u(t, 0, y) &= f_1(y) \\
u(t, a, y) &= f_2(y) \\
u(t, x, 0) &= f_3(x) \\
u(t, x, b) &= f_4(x) \\
u(0, x, y) &= g(x, y)
\end{aligned}
\tag{5.28}
$$

The problem can be broken down into five problems. $u = u_1 + u_2 + u_3 + u_4 + u_5$.

$$u_{1t} = u_{1xx} + u_{1yy}$$
$$u_1(0, x, y) = g(x, y) \tag{5.29}$$
$$u_1 = 0, \quad \text{all boundaries}$$

$$u_{2xx} + u_{2yy} = 0$$
$$u_2(0, y) = f_1(y) \tag{5.30}$$
$$u_2 = 0 \quad \text{on all other boundaries}$$

$$u_{3xx} + u_{3yy} = 0$$
$$u_3(a, y) = f_2(y) \tag{5.31}$$
$$u_3 = 0 \quad \text{on all other boundaries}$$

$$u_{4xx} + u_{4yy} = 0$$
$$u_4(x, 0) = f_3(x) \tag{5.32}$$
$$u_4 = 0 \quad \text{on all other boundaries}$$

$$u_{5xx} + u_{5yy} = 0$$
$$u_5(x, b) = f_4(x) \tag{5.33}$$
$$u_5 = 0 \quad \text{on all other boundaries}$$

5.1.1   Time-Dependent Boundary Conditions
We will explore this topic when we discuss Laplace transforms.

Example 5.4 (A finite cylinder).   Next we consider a cylinder of finite length $2L$ and radius r_1. As in the first problem in this chapter, there are two possible length scales and we choose r_1. The cylinder has temperature u_0 initially. The ends at $L = \pm L$ are suddenly insulated while the sides are exposed to a fluid at temperature u_1. The differential equation with no variation in the θ direction and the boundary conditions are

$$u_t = \frac{\alpha}{r}(r u_r)_r + u_{zz}$$
$$u_z(t, r, -L) = u_z(t, r, +L) = 0$$
$$k u_r(r_1) + h[u(r_1) - u_1(r_1)] = 0 \tag{5.34}$$
$$u(0, r, z) = u_0$$
$$u \text{ is bounded}$$

If we choose the length scale as r_1 then we define $\eta = r/r_1$, $\varsigma = z/L$, and $\tau = \alpha t/r_1^2$. The normalized temperature can be chosen as $U = (u - u_1)(u_0 - u_1)$. With these we find that

$$U_\tau = \frac{1}{\eta}(\eta U_\eta)_\eta + \left(\frac{r_1}{L}\right)^2 U_{\varsigma\varsigma}$$
$$U_\varsigma(\varsigma = \pm 1) = 0$$
$$U_\eta(\eta = 1) + BU(\eta = 1) = 0 \tag{5.35}$$
$$U(\tau = 0) = 1$$

where $B = hr_1/k$.

Let $U = T(\tau)R(\eta)Z(\varsigma)$. Insert into the differential equation and divide by U.

$$\frac{T'}{T} = \frac{1}{\eta R}(\eta R')' + \left(\frac{r_1}{L}\right)^2 \frac{Z''}{Z} \tag{5.36}$$

$$Z_\varsigma(\varsigma = \pm 1) = 0$$
$$R_\eta(\eta = 1) + BR(\eta = 1) = 0$$
$$U(\tau = 0) = 1$$

Again, the dance is the same. The left-hand side of Eq. (5.36) cannot be a function of η or ς so each side must be a constant. The constant must be negative for the time term to be bounded.

Experience tells us that Z''/Z must be a negative constant because otherwise Z would be exponential functions and we could not simultaneously satisfy the boundary conditions at $\varsigma = \pm 1$. Thus, we have

$$T' = -\lambda^2 T$$
$$\eta^2 R'' + \eta R' + \beta^2 \eta^2 R = 0 \tag{5.37}$$
$$Z'' = -\gamma^2 \left(\frac{L}{r_1}\right)^2 Z$$

with solutions

$$T = Ae^{-\lambda^2 t}$$
$$Z = C_1 \cos(\gamma L\varsigma/r_1) + C_2 \sin(\gamma L\varsigma/r_1) \tag{5.38}$$
$$R = C_3 J_0(\beta\eta) + C_4 Y_o(\beta\eta)$$

It is clear that C_4 must be zero always when the cylinder is not hollow because Y_0 is unbounded when $\eta = 0$. The boundary conditions at $\varsigma = \pm 1$ imply that Z is an even function, so that C_2

must be zero. The boundary condition at $\zeta = 1$ is

$$Z_\zeta = -C_1(\gamma L/r_1)\sin(\gamma L/r_1) = 0, \quad \text{or} \quad \gamma L/r_1 = n\pi \tag{5.39}$$

The boundary condition at $\eta = 1$ requires

$$C_3[J_0'(\beta) + BJ_0(\beta)] = 0 \text{ or}$$
$$BJ_0(\beta) = \beta J_1(\beta) \tag{5.40}$$

which is the transcendental equation for finding β_m. Also note that

$$\lambda^2 = \gamma_n^2 + \beta_m^2 \tag{5.41}$$

By superposition we write the final form of the solution as

$$U(\tau, \eta, \zeta) = \sum_{n=0}^{\infty}\sum_{m=0}^{\infty} K_{nm} e^{-(\gamma_n^2 + \beta_m^2)\tau} J_0(\beta_m \eta)\cos(n\pi\,\zeta) \tag{5.42}$$

K_{nm} is found using the orthogonality properties of $J_0(\beta_m\eta)$ and $\cos(n\pi\,\zeta)$ after using the initial condition.

$$\int_{r=0}^{1} r J_0(\beta_m\eta)d\eta \int_{\zeta=-1}^{1}\cos(n\pi\,\zeta)d\zeta = K_{nm}\int_{r=0}^{1} r J_0^2(\beta_m\eta)d\eta \int_{\zeta=-1}^{1}\cos^2(n\pi\,\zeta)d\zeta \tag{5.43}$$

Example 5.5 (Heat transfer in a sphere). Consider heat transfer in a solid sphere whose surface temperature is a function of θ, the angle measured downward from the z-axis (see Fig. 1.3, Chapter 1). The problem is steady and there is no heat source.

$$r\frac{\partial^2}{\partial r^2}(ru) + \frac{1}{\sin\theta}\frac{\partial}{\partial\theta}\left(\sin\theta\frac{\partial u}{\partial\theta}\right) = 0$$
$$u(r = 1) = f(\theta) \tag{5.44}$$
$$u \text{ is bounded}$$

Substituting $x = \cos\theta$,

$$r\frac{\partial^2}{\partial r^2}(ru) + \frac{\partial}{\partial x}\left[(1 - x^2)\frac{\partial u}{\partial x}\right] = 0 \tag{5.45}$$

We separate variables by assuming $u = R(r)X(x)$. Substitute into the equation, divide by RX and find

$$\frac{r}{R}(r)'' = -\frac{[(1 - x^2)X']'}{X} = \pm\lambda^2 \tag{5.46}$$

or

$$r(r R)'' \mp \lambda^2 R = 0$$

$$[(1 - x^2)X']' \pm \lambda^2 X = 0 \qquad (5.47)$$

The second of these is Legendre's equation, and we have seen that it has bounded solutions at $r = 1$ when $\pm\lambda^2 = n(n + 1)$. The first equation is of the Cauchy–Euler type with solution

$$R = C_1 r^n + C_2 r^{-n-1} \qquad (5.48)$$

Noting that the constant C_2 must be zero to obtain a bounded solution at $r = 0$, and using superposition,

$$u = \sum_{n=0}^{\infty} K_n r^n P_n(x) \qquad (5.49)$$

and using the condition at $f(r = 1)$ and the orthogonality of the Legendre polynomial

$$\int_{\theta=0}^{\pi} f(\theta) P_n(\cos\theta) d\theta = \int_{\theta=0}^{\pi} K_n P_n^2(\cos\theta) d\theta = \frac{2K_n}{2n + 1} \qquad (5.50)$$

$$K_n = \frac{2n + 1}{2} \int_{\theta=0}^{\pi} f(\theta) P_n(\cos\theta) d\theta \qquad (5.51)$$

5.2 VIBRATIONS PROBLEMS

We now consider some vibrations problems. In Chapter 2 we found a solution for a vibrating string initially displaced. We now consider the problem of a string forced by a sine function.

Example 5.6 (Resonance in a vibration problem). Equation (1.21) in Chapter 1 is

$$y_{tt} = a^2 y_{xx} + A \sin(\eta t) \qquad (5.52)$$

Select a length scale as L, the length of the string, and a time scale L/a and defining $\xi = x/L$ and $\tau = ta/L$,

$$y_{\tau\tau} = y_{\xi\xi} + C \sin(\omega\tau) \qquad (5.53)$$

where ω is a dimensionless frequency, $\eta L/a$ and $C = AL^2 a^2$.

The boundary conditions and initial velocity and displacement are all zero, so the boundary conditions are all homogeneous, while the differential equation is not. Back in Chapter 2 we saw one way of dealing with this. Note that it wouldn't have worked had q''' been a function of time. We approach this problem somewhat differently. From experience, we expect a solution of the form

$$y(\xi, \tau) = \sum_{n=1}^{\infty} B_n(\tau) \sin(n\pi\xi) \qquad (5.54)$$

where the coefficients $B_n(\tau)$ are to be determined. Note that the equation above satisfies the end conditions. Inserting this series into the differential equation and using the Fourier sine series of C

$$C = \sum_{n=1}^{\infty} \frac{2C[1 - (-1)^n]}{n\pi} \sin(n\pi\xi) \qquad (5.55)$$

$$\sum_{n=1}^{\infty} B_n''(\tau) \sin(n\pi\xi) = \sum_{n=1}^{\infty} [-(n\pi)^2 B_n(\tau)] \sin(n\pi\xi)$$

$$+ C \sum_{n=1}^{\infty} \frac{2[1 - (-1)^n]}{n\pi} \sin(n\pi\xi) \sin(\varpi\tau) \qquad (5.56)$$

Thus

$$B_n'' = -(n\pi)^2 B_n + C\frac{2[1 - (-1)^n]}{n\pi} \sin(\varpi\tau) \qquad (5.57)$$

subject to initial conditions $y = 0$ and $y_\tau = 0$ at $\tau = 0$. When n is even the solution is zero. That is, since the right-hand side is zero when n is even,

$$B_n = C_1 \cos(n\pi\tau) + C_2 \sin(n\pi\tau) \qquad (5.58)$$

But since both $B_n(0)$ and $B_n'(0)$ are zero, $C_1 = C_2 = 0$. When n is odd we can write

$$B_{2n-1}'' + [(2n - 1)\pi]^2 B_{2n-1} = \frac{4C}{(2n - 1)\pi} \sin(\omega\tau) \qquad (5.59)$$

$(2n - 1)\pi$ is the natural frequency of the system, ω_n. The homogeneous solution of the above equation is

$$B_{2n-1} = D_1 \cos(\omega_n\tau) + D_2 \sin(\omega_n\tau) . \qquad (5.60)$$

To obtain the particular solution we assume a solution in the form of sines and cosines.

$$B_P = E_1 \cos(\omega\tau) + E_2 \sin(\omega\tau) \qquad (5.61)$$

Differentiating and inserting into the differential equation we find

$$-E_1\omega^2 \cos(\omega\tau) - E_2\omega^2 \sin(\omega\tau) + \omega_n^2[E_1 \cos(\omega\tau) + E_2 \sin(\omega\tau)] = \frac{4C}{\omega_n} \sin(\omega\tau) \qquad (5.62)$$

Equating coefficients of sine and cosine terms

$$E_1(\omega_n^2 - \omega^2)\cos(\omega\tau) = 0 \qquad \omega \neq \omega_n$$

$$E_2(\omega_n^2 - \omega^2)\sin(\omega\tau) = \frac{4C}{\omega_n}\sin(\omega\tau) \tag{5.63}$$

Thus

$$E_1 = 0 \qquad E_2 = \frac{4C}{\omega_n(\omega_n^2 - \omega^2)} \qquad \omega \neq \omega_n \tag{5.64}$$

Combining the homogeneous and particular solutions

$$B_{2n-1} = D_1\cos(\omega_n\tau) + D_2\sin(\omega_n\tau) + \frac{4C}{\omega_n(\omega_n^2 - \omega^2)}\sin(\omega\tau) \tag{5.65}$$

The initial conditions at $\tau = 0$ require that

$$D_1 = 0$$

$$D_2 = -\frac{4C(\omega/\omega_n)}{\omega_n(\omega_n^2 - \omega^2)} \tag{5.66}$$

The solution for B_{2n-1} is

$$B_{2n-1} = \frac{4C}{\omega_n(\omega^2 - \omega_n^2)}\left(\frac{\omega}{\omega_n}\sin(\omega_n\tau) - \sin(\omega\tau)\right), \ \omega \neq \omega_n \tag{5.67}$$

The solution is therefore

$$y(\xi, \tau) = 4C\sum_{n=1}^{\infty}\frac{\sin(\omega_n\xi)}{\omega_n(\omega^2 - \omega_n^2)}\left(\frac{\omega}{\omega_n}\sin(\omega_n\tau) - \sin(\omega\tau)\right) \tag{5.68}$$

When $\omega = \omega_n$ the above is not valid. The form of the particular solution should be chosen as

$$B_P = E_1\tau\cos(\omega\tau) + E_2\tau\sin(\omega\tau) \tag{5.69}$$

Differentiating and inserting into the differential equation for B_{2n-1}

$$[E_1\tau\omega_n^2 + 2E_2\omega_n - E_1\tau\omega_n^2]\cos(\omega_n\tau) + [E_2\tau\omega_n^2 - E_2\tau\omega_n^2 - 2E_1\omega_n]\sin(\omega_n\tau) = \frac{4C}{\omega_n}\sin(\omega_n\tau)$$
$$\tag{5.70}$$

Thus

$$E_2 = 0 \qquad E_1 = -\frac{4C}{2\omega_n^2} \tag{5.71}$$

and the solution when $\omega = \omega_n$ is

$$B_{2n-1} = C_1 \cos(\omega_n \tau) + C_2 \sin(\omega_n \tau) - \frac{2C}{\omega_n^2} \tau \cos(\omega_n \tau) \qquad (5.72)$$

The initial condition on position implies that $C_1 = 0$. The initial condition that the initial velocity is zero gives

$$\omega_n C_2 - \frac{2C}{\omega_n^2} = 0 \qquad (5.73)$$

The solution for B_{2n-1} is

$$B_{2n-1} = \frac{2C}{\omega_n^3} [\sin(\omega_n \tau) - \omega_n \tau \cos(\omega_n \tau)] \qquad (5.74)$$

Superposition now gives

$$y(\xi, \tau) = \sum_{n=1}^{\infty} \frac{2C}{\omega_n^3} \sin(\omega_n \xi)[\sin(\omega_n \tau) - \omega_n \tau \cos(\omega_n \tau)] \qquad (5.75)$$

An interesting feature of the solution is that there are an infinite number of natural frequencies,

$$\eta = \frac{a}{L}[\pi, 3\pi, 5\pi, \ldots, (2n-1)\pi, \ldots] \qquad (5.76)$$

If the system is excited at *any* of the frequencies, the magnitude of the oscillation will grow (theoretically) without bound. The smaller natural frequencies will cause the growth to be fastest.

Example 5.7 (Vibration of a circular membrane). Consider now a circular membrane (like a drum). The partial differential equation describing the displacement $y(t, r, \theta)$ was derived in Chapter 1.

$$a^{-2} \frac{\partial^2 y}{\partial t^2} = \frac{1}{r} \frac{\partial}{\partial r} \left(r \frac{\partial y}{\partial r} \right) + \frac{1}{r^2} \frac{\partial^2 y}{\partial \theta^2} \qquad (5.77)$$

Suppose it has an initial displacement of $y(0, r, \theta) = f(r, \theta)$ and the velocity $y_t = 0$. The displacement at $r = r_1$ is also zero and the displacement must be finite for all r, θ, and t. The length scale is r_1 and the time scale is r_1/a. $r/r_1 = \eta$ and $ta/r_1 = \tau$.

We have

$$\frac{\partial^2 y}{\partial \tau^2} = \frac{1}{\eta} \frac{\partial}{\partial \eta} \left(\eta \frac{\partial y}{\partial \eta} \right) + \frac{1}{\eta^2} \frac{\partial^2 y}{\partial \theta^2} \qquad (5.78)$$

Separation of variables as $y = T(\tau)R(\eta)S(\theta)$, substituting into the equation and dividing by TRS,

$$\frac{T''}{T} = \frac{1}{\eta R}(\eta R')' + \frac{1}{\eta^2}\frac{S''}{S} = -\lambda^2 \tag{5.79}$$

The negative sign is because we anticipate sine and cosine solutions for T.

We also note that

$$\lambda^2\eta^2 + \frac{\eta}{R}(\eta R')' = -\frac{S''}{S} = \pm\beta^2 \tag{5.80}$$

To avoid exponential solutions in the θ direction we must choose the positive sign. Thus we have

$$T'' = -\lambda^2 T$$
$$S'' = -\beta^2 S \tag{5.81}$$
$$\eta(\eta R')' + (\eta^2\lambda^2 - \beta^2)R = 0$$

The solutions of the first two of these are

$$T = A_1\cos(\lambda\tau) + A_2\sin(\lambda\tau)$$
$$S = B_1\cos(\beta\theta) + B_2\sin(\beta\theta) \tag{5.82}$$

The boundary condition on the initial velocity guarantees that $A_2 = 0$. β must be an integer so that the solution comes around to the same place after θ goes from 0 to 2π. Either B_1 and B_2 can be chosen zero because it doesn't matter where θ begins (we can adjust $f(r, \theta)$).

$$T(\tau)S(\theta) = AB\cos(\lambda\tau)\sin(n\theta) \tag{5.83}$$

The differential equation for R should be recognized from our discussion of Bessel functions. The solution with $\beta = n$ is the Bessel function of the first kind order n. The Bessel function of the second kind may be omitted because it is unbounded at $r = 0$. The condition that $R(1) = 0$ means that λ is the mth root of $J_n(\lambda_{mn}) = 0$. The solution can now be completed using superposition and the orthogonality properties.

$$y(\tau, \eta, \theta) = \sum_{n=0}^{\infty}\sum_{m=1}^{\infty} K_{nm}J_n(\lambda_{mn}\eta)\cos(\lambda_{mn}\tau)\sin(n\theta) \tag{5.84}$$

Using the initial condition

$$f(\eta, \theta) = \sum_{n=0}^{\infty}\sum_{m=1}^{\infty} K_{nm}J_n(\lambda_{mn}\eta)\sin(n\theta) \tag{5.85}$$

and the orthogonality of $\sin(n\theta)$ and $J_n(\lambda_{mn}\eta)$

$$\int_{\theta=0}^{2\pi}\int_{\eta=0}^{1} f(\eta, \theta)\eta J_n(\lambda_{mn}\eta) \sin(n\theta)d\theta\,d\eta = K_{nm} \int_{\theta=0}^{2\pi} \sin^2(n\theta)d\theta \int_{r=0}^{1} \eta J_n^2(\lambda_{mn}\eta)d\eta$$

$$= \frac{K_{nm}}{4} J_{n+1}^2(\lambda_{mn})$$

(5.86)

$$K_{nm} = \frac{4}{J_{n+1}^2(\lambda_{nm})} \int_{\theta=0}^{2\pi}\int_{\eta=0}^{1} f(\eta, \theta)\eta J_n(\lambda_{nm}\eta) \sin(n\theta)d\theta\,d\eta$$

(5.87)

Problems

1.   The conduction equation in one dimension is to be solved subject to an insulated surface at $x = 0$ and a convective boundary condition at $x = L$. Initially the temperature is $u(0, x) = f(x)$, a function of position. Thus

$$u_t = \alpha\,u_{xx}$$
$$u_x(t, 0) = 0$$
$$ku_x(t, L) = -h[u(t, L) - u_1]$$
$$u(0, x) = f(x)$$

First nondimensionalize and normalize the equations. Then solve by separation of variables. Find a specific solution when $f(x) = 1 - x^2$.

2.   Consider the diffusion problem

$$u_t = \alpha\,u_{xx} + q(x)$$
$$u_x(t, 0) = 0$$
$$u_x(t, L) = -h[u(t, L) - u_1]$$
$$u(0, x) = u_1$$

Define time and length scales and define a u scale such that the initial value of the dependent variable is zero. Solve by separation of variables and find a specific solution for $q(x) = Q$, a constant. Refer to Problem 2.1 in Chapter 2.

3.   Solve the steady-state conduction

$$u_{xx} + u_{yy} = 0$$

$$u_x(0, y) = 0$$

$$u(a, y) = u_0$$

$$u(x, 0) = u_1$$

$$u_y(x, b) = -h[u(x, b) - u_1]$$

Note that one could choose a length scale either a or b. Choose a. Note that if you choose

$$U = \frac{u - u_1}{u_0 - u_1}$$

there is only one nonhomogeneous boundary condition and it is normalized. Solve by separation of variables.

5.3   FOURIER INTEGRALS

We consider now problems in which one dimension of the domain is infinite in extent. Recall that a function defined on an interval $(-c, c)$ can be represented as a Fourier series

$$f(x) = \frac{1}{2c} \int_{\varsigma=-c}^{c} f(\varsigma)d\varsigma + \frac{1}{c}\sum_{n=1}^{\infty} \int_{\varsigma=-c}^{c} f(\varsigma)\cos\left(\frac{n\pi\varsigma}{c}\right)d\varsigma \cos\left(\frac{n\pi x}{c}\right)$$

$$+ \frac{1}{c}\sum_{n=1}^{\infty} \int_{\varsigma=-c}^{c} f(\varsigma)\sin\left(\frac{n\pi\varsigma}{c}\right)d\varsigma \sin\left(\frac{n\pi x}{c}\right) \tag{5.88}$$

which can be expressed using trigonometric identities as

$$f(x) = \frac{1}{2c} \int_{\varsigma=-c}^{c} f(\varsigma)d\varsigma + \frac{1}{c}\sum_{n=1}^{\infty} \int_{\varsigma=-c}^{c} f(\varsigma)\cos\left[\frac{n\pi}{c}(\varsigma - x)\right]d\varsigma \tag{5.89}$$

We now formally let c approach infinity. If $\int_{\varsigma=-c}^{\infty} f(\varsigma)d\varsigma$ exists, the first term vanishes. Let $\Delta\alpha = \pi/c$. Then

$$f(x) = \frac{2}{\pi}\sum_{n=1}^{\infty} \int_{\varsigma=0}^{c} f(\varsigma)\cos[n\Delta\alpha(\varsigma - x)]d\varsigma\, \Delta\alpha \tag{5.90}$$

or, with

$$g_c(n\Delta\alpha, x) = \int\limits_{\varsigma=0}^{c} f(\varsigma)\cos[n\Delta\,\alpha(\varsigma - x)]d\varsigma \qquad (5.91)$$

we have

$$f(x) = \sum_{n=1}^{\infty} g_c(n\Delta\alpha, x)\Delta\alpha \qquad (5.92)$$

As c approaches infinity we can imagine that $\Delta\alpha$ approaches $d\alpha$ and $n\Delta\alpha$ approaches α, whereupon the equation for $f(x)$ becomes an integral expression

$$f(x) = \frac{2}{\pi} \int\limits_{\varsigma=0}^{\infty} \int\limits_{\alpha=0}^{\infty} f(\varsigma)\cos[\alpha(\varsigma - x)]d\varsigma\, d\alpha \qquad (5.93)$$

which can alternatively be written as

$$f(x) = \int\limits_{\alpha=0}^{\infty} [A(\alpha)\cos\alpha\,x + B(\alpha)\sin\alpha\,x]d\alpha \qquad (5.94)$$

where

$$A(\alpha) = \frac{2}{\pi} \int\limits_{\varsigma=0}^{\infty} f(\varsigma)\cos\alpha\varsigma\, d\varsigma \qquad (5.95)$$

and

$$B(\alpha) = \frac{2}{\pi} \int\limits_{\varsigma=0}^{\infty} f(\varsigma)\sin\alpha\varsigma\, d\varsigma \qquad (5.96)$$

Example 5.8 (Transient conduction in a semi-infinite region). Consider the boundary value problem

$$u_t = u_{xx} \qquad (x \geq 0,\ t \geq 0)$$
$$u(0, t) = 0 \qquad\qquad\qquad (5.97)$$
$$u(x, 0) = f(x)$$

This represents transient heat conduction with an initial temperature $f(x)$ and the boundary at $x = 0$ suddenly reduced to zero. Separation of variables as $T(t)X(x)$ would normally yield a solution of the form

$$B_n \exp(-\lambda^2 t) \sin\left(\frac{\lambda x}{c}\right) \tag{5.98}$$

for a region of x on the interval $(0, c)$. Thus, for x on the interval $0 \leq x \leq \infty$ we have

$$B(\alpha) = \frac{2}{\pi} \int\limits_{\varsigma=0}^{\infty} f(\varsigma) \sin \alpha \varsigma \, d\varsigma \tag{5.99}$$

and the solution is

$$u(x, t) = \frac{2}{\pi} \int\limits_{\lambda=0}^{\infty} \exp(-\lambda^2 t) \sin(\lambda x) \int\limits_{s=0}^{\infty} f(s) \sin(\lambda s) ds \, d\alpha \tag{5.100}$$

Noting that

$$2 \sin \alpha s \, \sin \alpha x = \cos \alpha \, (s - x) - \cos \alpha (s + x) \tag{5.101}$$

and that

$$\int\limits_0^{\infty} \exp(-\gamma^2 \alpha) \cos(\gamma b) d\gamma = \frac{1}{2} \sqrt{\frac{\pi}{\alpha}} \exp\left(-\frac{b^2}{4\alpha}\right) \tag{5.102}$$

we have

$$u(x, t) = \frac{1}{2\sqrt{\pi t}} \int\limits_0^{\infty} f(s) \left\{ \exp\left[-\frac{(s - x)^2}{4t}\right] - \exp\left[-\frac{(s + x)^2}{4t}\right] \right\} ds \tag{5.103}$$

Substituting into the first of these integrals $\sigma^2 = \frac{(s-x)^2}{4t}$ and into the second integral

$$\sigma^2 = \frac{(s + x)^2}{4t} \tag{5.104}$$

$$u(x, t) = \frac{1}{\sqrt{\pi}} \int\limits_{-x/2\sqrt{t}}^{\infty} f(x + 2\sigma \sqrt{t}) e^{-\sigma^2} d\sigma$$

$$- \frac{1}{\sqrt{\pi}} \int\limits_{x/2\sqrt{t}}^{\infty} f(-x + 2\sigma \sqrt{t}) e^{-\sigma^2} d\sigma \tag{5.105}$$

In the special case where $f(x) = u_0$

$$u(x, t) = \frac{2u_0}{\sqrt{\pi}} \int_0^{x/2\sqrt{t}} \exp(-\sigma^2)d\sigma = u_0 \operatorname{erf}\left(\frac{x}{2\sqrt{t}}\right) \qquad (5.106)$$

where $\operatorname{erf}(p)$ is the Gauss error function defined as

$$\operatorname{erf}(p) = \frac{2}{\sqrt{\pi}} \int_0^p \exp(-\sigma^2)d\sigma \qquad (5.107)$$

Example 5.9 (Steady conduction in a quadrant).   Next we consider steady conduction in the region $x \geq 0$, $y \geq 0$ in which the face at $x = 0$ is kept at zero temperature and the face at $y = 0$ is a function of x: $u = f(x)$. The solution is also assumed to be bounded.

$$u_{xx} + u_{yy} = 0 \qquad (5.108)$$

$$u(x, 0) = f(x) \qquad (5.109)$$

$$u(0, y) = 0 \qquad (5.110)$$

Since $u(0, y) = 0$ the solution should take the form $e^{-\alpha y} \sin \alpha x$, which is, according to our experience with separation of variables, a solution of the equation $\nabla^2 u = 0$. We therefore assume a solution of the form

$$u(x, y) = \int_0^\infty B(\alpha)e^{-\alpha y} \sin \alpha x d\alpha \qquad (5.111)$$

with

$$B(\alpha) = \frac{2}{\pi} \int_0^\infty f(\varsigma) \sin \alpha \varsigma \, d\varsigma \qquad (5.112)$$

The solution can then be written as

$$u(x, y) = \frac{2}{\pi} \int_{\varsigma=0}^\infty f(\varsigma) \int_{\alpha=0}^\infty e^{-\alpha y} \sin \alpha x \sin \alpha \varsigma \, d\alpha \, d\varsigma \qquad (5.113)$$

Using the trigonometric identity for $2 \sin ax \sin a\varsigma = \cos a(\varsigma - x) - \cos a(\varsigma + x)$ and noting that

$$\int_0^\infty e^{-\alpha y} \cos a\beta \, d\alpha = \frac{y}{\beta^2 + y^2} \qquad (5.114)$$

we find

$$u(x, y) = \frac{y}{\pi} \int_0^\infty f(\varsigma) \left[\frac{1}{(\varsigma - x)^2 + y^2} - \frac{1}{(\varsigma + x)^2 + y^2} \right] d\varsigma \qquad (5.115)$$

Problem

Consider the transient heat conduction problem

$$u_t = u_{xx} + u_{yy} \quad x \geq 0, \ 0 \leq y \leq 1, \quad t \geq 0$$

with boundary and initial conditions

$$u(t, 0, y) = 0$$
$$u(t, x, 0) = 0$$
$$u(t, x, 1) = 0$$
$$u(0, x, y) = u_0$$

and $u(t, x, y)$ is bounded.

Separate the problem into two problems $u(t, x, y) = v(t, x)w(t, y)$ and give appropriate boundary conditions. Show that the solution is given by

$$u(t, x, y) = \frac{4}{\pi} \mathrm{erf}\left[\frac{x}{2\sqrt{t}} \right] \sum_{n=1}^\infty \frac{\sin(2n-1)\pi y}{2n-1} \exp[-(2n-1)^2 \pi^2 t]$$

FURTHER READING

V. Arpaci, *Conduction Heat Transfer*. Reading, MA: Addison-Wesley, 1966.

J. W. Brown and R. V. Churchill, *Fourier Series and Boundary Value Problems*. 6th edition. New York: McGraw-Hill, 2001.

CHAPTER 6

Integral Transforms: The Laplace Transform

Integral transforms are a powerful method of obtaining solutions to both ordinary and partial differential equations. They are used to change ordinary differential equations into algebraic equations and partial differential into ordinary differential equations. The general idea is to multiply a function $f(t)$ of some independent variable t (not necessarily time) by a Kernel function $K(t, s)$ and integrate over some t space to obtain a function $F(s)$ of s which one hopes is easier to solve. Of course one must then inverse the process to find the desired function $f(t)$. In general,

$$F(s) = \int_{t=a}^{b} K(t, s) f(t) dt \tag{6.1}$$

6.1   THE LAPLACE TRANSFORM
A useful and widely used integral transform is the Laplace transform, defined as

$$L[f(t)] = F(s) = \int_{t=0}^{\infty} f(t) e^{-st} dt \tag{6.2}$$

Obviously, the integral must exist. The function $f(t)$ must be sectionally continuous and of exponential order, which is to say $|f(t)| \leq Me^{kt}$ when $t > 0$ for some constants M and k. For example neither the Laplace transform of t^{-1} nor $\exp(t^2)$ exists.

The inversion formula is

$$L^{-1}[F(s)] = f(t) = \frac{1}{2\pi i} \lim_{L \to \infty} \int_{\gamma-iL}^{\gamma+iL} F(s) e^{ts} ds \tag{6.3}$$

in which $\gamma - iL$ and $\gamma + iL$ are complex numbers. We will put off using the inversion integral until we cover complex variables. Meanwhile, there are many tables giving Laplace transforms and inverses. We will now spend considerable time developing the theory.

6.2   SOME IMPORTANT TRANSFORMS

6.2.1   Exponentials

First consider the exponential function:

$$L[e^{-at}] = \int_{t=0}^{\infty} e^{-at} e^{-st} dt = \int_{t=0}^{\infty} e^{-(s=a)t} dt = \frac{1}{s+a} \tag{6.4}$$

If $a = 0$, this reduces to

$$L[1] = 1/s \tag{6.5}$$

6.2.2   Shifting in the s-domain

$$L[e^{at} f(t)] = \int_{t=0}^{\infty} e^{-(s-a)t} f(t) dt = F(s-a) \tag{6.6}$$

6.2.3   Shifting in the time domain

Consider a function defined as

$$f(t) = 0 \quad t < a \qquad f(t) = f(t-a) \quad t > a \tag{6.7}$$

Then

$$\int_{\tau=0}^{\infty} e^{-s\tau} f(\tau - a) d\tau = \int_{\tau=0}^{a} 0 d\tau + \int_{\tau=a}^{\infty} e^{-s\tau} f(\tau - a) d\tau \tag{6.8}$$

Let $\tau - a = t$. Then

$$\int_{t=0}^{\infty} e^{-s(t+a)} f(t) dt = F(s)e^{-as} = L[f(t-a)] \tag{6.9}$$

the shifted function described above.

6.2.4 Sine and cosine

Now consider the sine and cosine functions. We shall see in the next chapter (and you should already know) that

$$e^{ikt} = \cos(kt) + i\sin(kt) \tag{6.10}$$

Thus the Laplace transform is

$$L[e^{ikt}] = L[\cos(kt)] + iL[\sin(kt)] = \frac{1}{s - ik} = \frac{s + ik}{(s + ik)(s - ik)} = \frac{s}{s^2 + k^2} + i\frac{k}{s^2 + k^2} \tag{6.11}$$

so

$$L[\sin(kt)] = \frac{k}{s^2 + k^2} \tag{6.12}$$

$$L[\cos(kt)] = \frac{s}{s^2 + k^2} \tag{6.13}$$

6.2.5 Hyperbolic functions

Similarly for hyperbolic functions

$$L[\sinh(kt)] = L\left[\frac{1}{2}(e^{kt} - e^{-kt})\right] = \frac{1}{2'}\left[\frac{1}{s - k} - \frac{1}{s + k}\right] = \frac{k}{s^2 - k^2} \tag{6.14}$$

Similarly,

$$L[\cosh(kt)] = \frac{s}{s^2 - k^2} \tag{6.15}$$

6.2.6 Powers of t: t^m

We shall soon see that the Laplace transform of t^m is

$$L[t^m] = \frac{\Gamma(m + 1)}{s^{m+1}} \qquad m > -1 \tag{6.16}$$

Using this together with the s domain shifting results,

$$L[t^m e^{-at}] = \frac{\Gamma(m + 1)}{(s + a)^{m+1}} \tag{6.17}$$

Example 6.1. Find the inverse transform of the function

$$F(s) = \frac{1}{(s - 1)^3}$$

This is a function that is shifted in the s-domain and hence Eq. (6.6) is applicable. Noting that $L^{-1}(1/s^3) = t^2/\Gamma(3) = t^2/2$ from Eq. (6.16)

$$f(t) = \frac{t^2}{2}e^t$$

Or we could use Eq. (6.17) directly.

Example 6.2.   Find the inverse transform of the function

$$F(s) = \frac{3}{s^2 + 4}e^{-s}$$

The inverse transform of

$$F(s) = \frac{2}{s^2 + 4}$$

is, according to Eq. (6.11)

$$f(t) = \frac{3}{2}\sin(2t)$$

The exponential term implies shifting in the time domain by 1. Thus

$$f(t) = 0, \quad t < 1$$
$$= \frac{3}{2}\sin[2(t-1)], \quad t > 1$$

Example 6.3.   Find the inverse transform of

$$F(s) = \frac{s}{(s-2)^2 + 1}$$

The denominator is shifted in the s-domain. Thus we shift the numerator term and write $F(s)$ as two terms

$$F(s) = \frac{s-2}{(s-2)^2 + 1} + \frac{2}{(s-2)^2 + 1}$$

Equations (6.6), (6.12), and (6.13) are applicable. The inverse transform of the first of these is a shifted cosine and the second is a shifted sine. Therefore each must be multiplied by exp(2t). The inverse transform is

$$f(t) = e^{2t}\cos(t) + 2e^{2t}\sin(t)$$

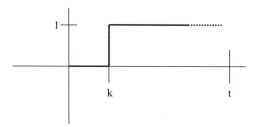

FIGURE 6.1: The Heaviside step

6.2.7   Heaviside step

A frequently useful function is the Heaviside step function, defined as

$$U_k(t) = 0 \quad 0 < t < k$$
$$= 1 \quad k < t \qquad (6.18)$$

It is shown in Fig. 6.1.
The Laplace transform is

$$L[U_k(t)] = \int\limits_{t=k}^{\infty} e^{-st} dt = \frac{1}{s} e^{-ks} \qquad (6.19)$$

The Heaviside step (sometimes called the unit step) is useful for finding the Laplace transforms of periodic functions.

Example 6.4 (Periodic functions).   For example, consider the periodic function shown in Fig. 6.2.

It can be represented by an infinite series of shifted Heaviside functions as follows:

$$f(t) = U_0 - 2U_k + 2U_{2k} - 2U_{3k} + \cdots = U_0 + \sum_{n=1}^{\infty} (-1)^n 2U_{nk} \qquad (6.20)$$

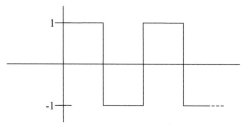

FIGURE 6.2: A periodic square wave

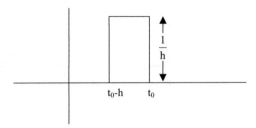

FIGURE 6.3: The Dirac delta function

The Laplace transform is found term by term,

$$L[f(t)] = \frac{1}{s}\{1 - 2e^{-sk}[1 - e^{-sk} + e^{-2sk} - e^{-3sk} \cdots]\}$$

$$= \frac{1}{s}\left\{1 - \frac{2e^{-sk}}{1 + e^{-sk}}\right\} = \frac{1}{s}\left(\frac{1 - e^{-sk}}{1 + e^{-sk}}\right) \qquad (6.21)$$

6.2.8   The Dirac delta function

Consider a function defined by

$$\lim \frac{U_{t_0} - U_{t_0-h}}{h} = \delta(t_0) \qquad h \to 0 \qquad (6.22)$$

$$L[\delta(t_0)] = e^{-st_0} \qquad (6.23)$$

The function, without taking limits, is shown in Fig. 6.3.

6.2.9   Transforms of derivatives

$$L\left[\frac{df}{dt}\right] = \int_{t=0}^{\infty} \frac{df}{dt} e^{-st} dt = \int_{t=0}^{\infty} e^{-st} df \qquad (6.24)$$

and integrating by parts

$$L\left[\frac{df}{dt}\right] = f(t)e^{-st}\Big|_0^{\infty} + s \int_{t=0}^{\infty} f(t)e^{-st} dt = s F(s) - f(0)$$

To find the Laplace transform of the second derivative we let $g(t) - f'(t)$. Taking the Laplace transform

$$L[g'(t)] = s G(s) - g(0)$$

and with

$$G(s) = L[f'(t)] = s\,F(s) - f(0)$$

we find that

$$L\left[\frac{d^2 f}{dt^2}\right] = s^2 F(s) - s f(0) - f'(0) \tag{6.25}$$

In general

$$L\left(\frac{d^n f}{dt^n}\right) = s^n F(s) - s^{n-1} f(0) - s^{n-2} f'(0) - \cdots - \frac{d^{n-1} f}{dt^{n-1}}(0) \tag{6.26}$$

The Laplace transform of t^m may be found by using the gamma function,

$$L[t^m] = \int_0^\infty t^m e^{-st}\,dt \quad\text{and}\quad \text{let } x = st \tag{6.27}$$

$$L[t^m] = \int_{x=0}^\infty \left(\frac{x}{s}\right)^m e^{-x}\frac{dx}{s} = \frac{1}{s^{m+1}} \int_{x=0}^\infty x^m e^{-x}\,dx = \frac{\Gamma(m+1)}{s^{m+1}} \tag{6.28}$$

which is true for all $m > -1$ even for nonintegers.

6.2.10 Laplace Transforms of Integrals

$$L\left[\int_{\tau=0}^t f(\tau)\,d\tau\right] = L[g(t)] \tag{6.29}$$

where $dg/dt = f(t)$. Thus $L[dg/dt] = s\,L[g(t)]$. Hence

$$L\left[\int_{\tau=0}^t f(\tau)\,d\tau\right] = \frac{1}{s}F(s) \tag{6.30}$$

6.2.11 Derivatives of Transforms

$$F(s) = \int_{t=0}^\infty f(t)e^{-st}\,dt \tag{6.31}$$

so

$$\frac{dF}{ds} = - \int_{t=0}^{\infty} tf(t)e^{-st}dt \tag{6.32}$$

and in general

$$\frac{d^n F}{ds^n} = L[(-t)^n f(t)] \tag{6.33}$$

For example

$$L[t \sin(kt)] = -\frac{d}{ds}\left(\frac{k}{s^2 + k^2}\right) = \frac{2sk}{(s^2 + k^2)^2} \tag{6.34}$$

6.3   LINEAR ORDINARY DIFFERENTIAL EQUATIONS WITH CONSTANT COEFFICIENTS

Example 6.5.   A homogeneous linear ordinary differential equation
Consider the differential equation

$$y'' + 4y' + 3y = 0$$
$$y(0) = 0 \tag{6.35}$$
$$y'(0) = 2$$

$$L[y''] = s^2 Y - s y(0) - y'(0) = s^2 Y - 2 \tag{6.36}$$

$$L[y'] = s Y - y(0) = s Y \tag{6.37}$$

Therefore

$$(s^2 + 4s + 3)Y = 2 \tag{6.38}$$

$$Y = \frac{2}{(s + 1)(s + 3)} = \frac{A}{s + 1} + \frac{B}{s + 3} \tag{6.39}$$

To solve for A and B, note that clearing fractions,

$$\frac{A(s + 3) + B(s + 1)}{(s + 1)(s + 3)} = \frac{2}{(s + 1)(s + 3)} \tag{6.40}$$

Equating the numerators, or

$$A + B = 0 \quad 3A + B = 2 : \quad A = 1 \quad B = -1 \tag{6.41}$$

and from Eq. (6.8)

$$Y = \frac{1}{s+1} - \frac{1}{s+3}$$
$$y = e^{-t} - e^{-3t}$$

(6.42)

6.4 SOME IMPORTANT THEOREMS
6.4.1 Initial Value Theorem

$$\lim_{s \to \infty} \int_{t=0}^{\infty} f'(t)e^{-st}dt = s\,F(s) - f(0) = 0$$

(6.43)

Thus

$$\lim_{s \to \infty} s\,F(s) = \lim_{t \to 0} f(t)$$

(6.44)

6.4.2 Final Value Theorem
As s approaches zero the above integral approaches the limit as t approaches infinity minus $f(0)$. Thus

$$\lim_{s \to 0} s\,F(s) = \lim_{t \to \infty} f(t)$$

(6.45)

6.4.3 Convolution
A very important property of Laplace transforms is the convolution integral. As we shall see later, it allows us to write down solutions for very general forcing functions and also, in the case of partial differential equations, to treat both time dependent forcing and time dependent boundary conditions.

Consider the two functions $f(t)$ and $g(t)$. $F(s) = L[f(t)]$ and $G(s) = L[g(t)]$. Because of the time shifting feature,

$$e^{-s\tau}G(s) = L[g(t-\tau)] = \int_{t=0}^{\infty} e^{-st}g(t-\tau)dt$$

(6.46)

$$F(s)G(s) = \int_{\tau=0}^{\infty} f(\tau)e^{-s\tau}G(s)d\tau$$

(6.47)

But

$$e^{-s\tau} G(s) = \int_{t=0}^{\infty} e^{-st} g(t - \tau) dt \qquad (6.48)$$

so that

$$F(s)G(s) = \int_{t=0}^{\infty} e^{-st} \int_{\tau=0}^{t} f(\tau)g(t - \tau)d\tau \, dt \qquad (6.49)$$

where we have used the fact that $g(t - \tau) = 0$ when $\tau > t$. The inverse transform of $F(s)G(s)$ is

$$L^{-1}[F(s)G(s)] = \int_{\tau=0}^{t} f(\tau)g(t - \tau)d\tau \qquad (6.50)$$

6.5   PARTIAL FRACTIONS

In the example differential equation above we determined two roots of the polynomial in the denominator, then separated the two roots so that the two expressions could be inverted in forms that we already knew. The method of separating out the expressions $1/(s + 1)$ and $1/(s + 3)$ is known as the method of partial fractions. We now develop the method into a more user friendly form.

6.5.1   Nonrepeating Roots

Suppose we wish to invert the transform $F(s) = p(s)/q(s)$, where $p(s)$ and $q(s)$ are polynomials. We first note that the inverse exists if the degree of $p(s)$ is lower than that of $q(s)$. Suppose $q(s)$ can be factored and a *nonrepeated root* is a.

$$F(s) = \frac{\phi(s)}{s - a} \qquad (6.51)$$

According to the theory of partial fractions there exists a constant C such that

$$\frac{\phi(s)}{s - a} = \frac{C}{s - a} + H(s) \qquad (6.52)$$

Multiply both sides by $(s - a)$ and take the limit as $s \to a$ and the result is

$$C = \phi(a) \qquad (6.53)$$

Note also that the limit of

$$p(s)\frac{s-a}{q(s)} \qquad (6.54)$$

as s approaches a is simply $p(s)/q'(s)$.

If $q(s)$ has no repeated roots and is of the form

$$q(s) = (s - a_1)(s - a_2)(s - a_3) \cdots (s - a_n) \qquad (6.55)$$

then

$$L^{-1}\left[\frac{p(s)}{q(s)}\right] = \sum_{m=1}^{n} \frac{p(a_m)}{q'(a_m)}e^{a_m t} \qquad (6.56)$$

Example 6.6. Find the inverse transform of

$$F(s) = \frac{4s + 1}{(s^2 + s)(4s^2 - 1)}$$

First separate out the roots of $q(s)$

$$q(s) = 4s(s + 1)(s + 1/2)(s - 1/2)$$
$$q(s) = 4s^4 + 4s^3 - s^2 - s$$
$$q'(s) = 16s^3 + 12s^2 - 2s - 1$$

Thus

$$q'(0) = -1 \qquad p(0) = 1$$
$$q'(-1) = -3 \qquad p(-1) = -3$$
$$q'(-1/2) = 1 \qquad p(-1/2) = -1$$
$$q'(1/2) = 3 \qquad p(1/2) = 3$$
$$f(t) = e^{-t} - e^{-t/2} + e^{t/2} - 1$$

Example 6.7. Solve the differential equation

$$y'' - y = 1 - e^{3t}$$

subject to initial conditions

$$y'(0) = y(0) = 0$$

Taking the Laplace transform

$$(s^2 - 1)Y = \frac{1}{s} - \frac{1}{s-3}$$

$$Y(s) = \frac{1}{s(s^2-1)} - \frac{1}{(s-3)(s^2-1)} = \frac{1}{s(s+1)(s-1)} - \frac{1}{(s-3)(s+1)(s-1)}$$

First find the inverse transform of the first term.

$$q = s^3 - s$$
$$q' = 3s^2 - 1$$
$$q'(0) = -1 \qquad p(0) = 1$$
$$q'(1) = 2 \qquad p(1) = 1$$
$$q'(-1) = 2 \qquad p(-1) = 1$$

The inverse transform is

$$-1 + 1/2e^t + 1/2\,e^{-t}$$

Next consider the second term.

$$q = s^3 - 3s^2 - s + 3$$
$$q' = 3s^2 - 6s - 1$$
$$q'(-3) = 44 \qquad p(-3) = 1$$
$$q'(1) = -4 \qquad p(1) = 1$$
$$q'(-1) = 8 \qquad p(-1) = 1$$

The inverse transform is

$$\frac{1}{44}e^{-3t} - \frac{1}{4}e^t + \frac{1}{8}e^{-t}$$

Thus

$$y(t) = \frac{1}{4}e^t + \frac{5}{8}e^{-t} + \frac{1}{44}e^{-3t} - 1$$

6.5.2   Repeated Roots

We now consider the case when $q(s)$ has a repeated root $(s + a)^{n+1}$. Then

$$F(s) = \frac{p(s)}{q(s)} = \frac{\phi(s)}{(s-a)^{n+1}} \qquad n = 1, 2, 3, \ldots$$

$$= \frac{A_a}{(s-a)} + \frac{A_1}{(s-a)^2} + \cdots + \frac{A_n}{(s-a)^{n+1}} + H(s) \qquad (6.57)$$

It follows that

$$\phi(s) = A_0(s-a)^n + \cdots + A_m(s-a)^{n-m} + \cdots + A_n + (s-a)^{n+1}H(s) \qquad (6.58)$$

By letting $s \to a$ we see that $A_n = \phi(a)$. To find the remaining A's, differentiate ϕ $(n-r)$ times and take the limit as $s \to a$.

$$\phi^{(n-r)}(a) = (n-r)!A_r \qquad (6.59)$$

Thus

$$F(s) = \sum_{r=0}^{n} \frac{\phi^{(n-r)}(a)}{(n-r)!} \frac{1}{(s-a)^{r+1}} + H(s) \qquad (6.60)$$

If the inverse transform of $H(s)$ (the part containing no repeated roots) is $h(t)$ it follows from the shifting theorem and the inverse transform of $1/s^m$ that

$$f(t) = \sum_{r=0}^{n} \frac{\phi^{(n-r)}(a)}{(n-r)!r!} t^r e^{at} + h(t) \qquad (6.61)$$

Example 6.8.   Inverse transform with repeated roots

$$F(s) = \frac{s}{(s+2)^3(s+1)} = \frac{A_0}{(s+2)} + \frac{A_1}{(s+2)^2} + \frac{A_2}{(s+2)^3} + \frac{C}{(s+1)}$$

Multiply by $(s+2)^3$.

$$\frac{s}{(s+1)} = A_0(s+2)^2 + A_1(s+2) + A_2 + \frac{C(s+2)^3}{(s+1)} = \phi(s)$$

Take the limit as $s \to -2$,

$$A_2 = 2$$

Differentiate once

$$\phi' = \frac{1}{(s+1)^2} \qquad \phi'(-2) = 1 = A_1$$

$$\phi'' = \frac{-2}{(s+1)^3} \qquad \phi''(-2) = 2 = A_0$$

To find C, multiply by $(s+1)$ and take $s = -1$ (in the original equation).

$$C = -1.$$

Thus

$$F(s) = \frac{2}{(s+2)} + \frac{1}{(s+2)^2} + \frac{2}{(s+2)^3} - \frac{1}{(s+1)}$$

and noting the shifting theorem and the theorem on t^m,

$$f(t) = 2e^{-2t} + te^{-2t} + 2t^2e^{-2t} + e^{-t}$$

6.5.3   Quadratic Factors: Complex Roots

If $q(s)$ has complex roots and all the coefficients are real this part of $q(s)$ can always be written in the form

$$(s-a)^2 + b^2 \tag{6.62}$$

This is a shifted form of

$$s^2 + b^2 \tag{6.63}$$

This factor in the denominator leads to sines or cosines.

Example 6.9.   Quadratic factors

Find the inverse transform of

$$F(s) = \frac{2(s-1)}{s^2 + 2s + 5} = \frac{2s}{(s+1)^2 + 4} - \frac{1}{(s+1)^2 + 4}$$

Because of the shifted s in the denominator the numerator of the first term must also be shifted to be consistent. Thus we rewrite as

$$F(s) = \frac{2(s+1)}{(s+1)^2 + 4} - \frac{3}{(s+1)^2 + 4}$$

The inverse transform of

$$\frac{2s}{s^2 + 4}$$

is

$$2\cos(2t)$$

and the inverse of

$$\frac{-3}{s^2+4} = -\frac{3}{2}\frac{2}{(s^2+4)}$$

is

$$-\frac{3}{2}\sin(2t)$$

Thus

$$f(t) = 2e^{-t}\cos(2t) - \frac{3}{2}e^{-t}\sin(2t)$$

Tables of Laplace transforms and inverse transforms can be found in many books such as the book by Arpaci and in the Schaum's Outline referenced below. A brief table is given here in Appendix A.

Problems

1. Solve the problem

$$y''' - 2y'' + 5y' = 0$$
$$y(0) = y'(0) = 0 \qquad y''(0) = 1$$

using Laplace transforms.

2. Find the general solution using Laplace transforms

$$y'' + k^2 y = a$$

3. Use convolution to find the solution to the following problem for general $g(t)$. Then find the solution for $g(t) = t^2$.

$$y'' + 2y' + y = g(t)$$
$$y'(0) = y(0) = 0$$

4. Find the inverse transforms.

   (a) $$F(s) = \frac{s+c}{(s+a)(s+b)^2}$$

   (b) $$F(s) = \frac{1}{(s^2+a^2)s^3}$$

(c)
$$F(s) = \frac{(s^2 - a^2)}{(s^2 + a^2)^2}$$

5. Find the periodic function whose Laplace transform is

$$F(s) = \frac{1}{s^2}\left[\frac{1 - e^{-s}}{1 + e^{-s}}\right]$$

and plot your results for $f(t)$ for several periods.

FURTHER READING

M. Abramowitz and I. A. Stegun, Eds., *Handbook of Mathematical Functions with Formulas, Graphs, and Mathematical Tables*. New York: Dover Publications, 1974.

V. S. Arpaci, *Conduction Heat Transfer*. Reading, MA: Addison-Wesley, 1966.

R. V. Churchill, *Operational Mathematics*, 3rd edition. New York: McGraw-Hill, 1972.

I. H. Sneddon, *The Use of Integral Transforms*. New York: McGraw-Hill, 1972.

CHAPTER 7

Complex Variables and the Laplace Inversion Integral

7.1   BASIC PROPERTIES

A *complex number* z can be defined as an ordered pair of real numbers, say x and y, where x is the real part of z and y is the real value of the imaginary part:

$$z = x + iy \tag{7.1}$$

where $i = \sqrt{-1}$

I am going to assume that the reader is familiar with the elementary properties of addition, subtraction, multiplication, etc. In general, complex numbers obey the same rules as real numbers. For example

$$(x_1 + iy_1)(x_2 + iy_2) = x_1 x_2 - y_1 y_2 + i(x_1 y_2 + x_2 y_1) \tag{7.2}$$

The *conjugate* of z is

$$\bar{z} = x - iy \tag{7.3}$$

It is often convenient to represent complex numbers on Cartesian coordinates with x and y as the axes. In such a case, we can represent the complex number (or variable) z as

$$z = x + iy = r(\cos\theta + i\sin\theta) \tag{7.4}$$

as shown in Fig. 7.1. We also define the exponential function of a complex number as $\cos\theta + i\sin\theta = e^{i\theta}$ which is suggested by replacing x in series $e^x = \sum_{n=0}^{\infty} \frac{x^n}{n!}$ by $i\theta$.

Accordingly,

$$e^{i\theta} = \cos\theta + i\sin\theta \tag{7.5}$$

and

$$e^{-i\theta} = \cos\theta - i\sin\theta \tag{7.6}$$

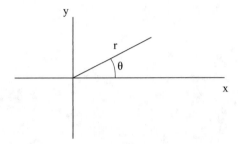

FIGURE 7.1: Polar representation of a complex variable z

Addition gives

$$\cos\theta = \frac{e^{i\theta} + e^{-i\theta}}{2} = \cosh(i\theta) \qquad (7.7)$$

and subtraction gives

$$\sin\theta = \frac{e^{i\theta} - e^{-i\theta}}{2i} = -i\sinh(i\theta) \qquad (7.8)$$

Note that

$$\cosh z = \frac{1}{2}\left(e^{x+iy} + e^{-x-iy}\right) = \frac{1}{2}\left(e^x[\cos y + i\sin y] + e^{-x}[\cos y - i\sin y]\right)$$

$$= \frac{e^x + e^{-x}}{2}\cos y + i\frac{e^x - e^{-x}}{2}\sin y$$

$$= \cosh x\cos y + i\sinh x\sin y \qquad (7.9)$$

The reader may show that

$$\sinh z = \sinh x\cos y + i\cosh x\sin y. \qquad (7.10)$$

Trigonometric functions are defined in the usual way:

$$\sin z = \frac{e^{iz} - e^{-iz}}{2i} \qquad \cos z = \frac{e^{iz} + e^{-iz}}{2} \qquad \tan z = \frac{\sin z}{\cos z} \qquad (7.11)$$

Two complex numbers are equal if and only if their real parts are equal and their imaginary parts are equal.

Noting that

$$z^2 = r^2(\cos^2\theta - \sin^2\theta + i2\sin\theta\cos\theta)$$
$$= r^2\left[\frac{1}{2}(1 + \cos 2\theta) - \frac{1}{2}(1 - \cos 2\theta) + i\sin 2\theta\right]$$
$$= r^2[\cos 2\theta + i\sin 2\theta]$$

We deduce that

$$z^{1/2} = r^{1/2}(\cos\theta/2 + i\sin\theta/2) \tag{7.12}$$

In fact in general

$$z^{m/n} = r^{m/n}[\cos(m\theta/n) + i\sin(m\theta/n)] \tag{7.13}$$

Example 7.1. Find $i^{1/2}$.

Noting that when $z = I$, $r = 1$ and $\theta = \pi/2$, with $m = 1$ and $n = 2$. Thus

$$i^{1/2} = 1^{1/2}[\cos(\pi/4) + i\sin(\pi/4)] = \frac{1}{\sqrt{2}}(1 + i)$$

Note, however, that if

$$w = \cos\left(\frac{\pi}{4} + \pi\right) + i\sin\left(\frac{\pi}{4} + \pi\right)$$

then $w^2 = i$. Hence $\frac{1}{\sqrt{2}}(-1 - i)$ is also a solution. The roots are shown in Fig. 7.2.

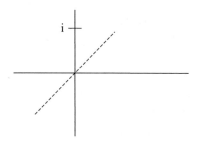

FIGURE 7.2: Roots of $i^{1/2}$

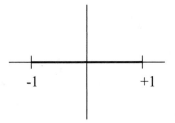

FIGURE 7.3: The roots of $1^{1/2}$

In fact in this example θ is also $\pi/2 + 2k\pi$. Using the fact that

$$z = re^{-i(\theta + 2k\pi)} \qquad k = 1, 2, 3, \ldots$$

it is easy to show that

$$z^{1/n} = \sqrt[n]{r}\left[\cos\left(\frac{\theta + 2\pi k}{n}\right) + i\sin\left(\frac{\theta + 2\pi k}{n}\right)\right] \qquad (7.14)$$

This is *De Moivre's theorem*. For example when $n = 2$ there are two solutions and when $n = 3$ there are three solutions. These solutions are called *branches* of $z^{1/n}$. A region in which the function is single valued is indicated by forming a *branch cut*, which is a line stretching from the origin outward such that the region between the positive real axis and the line contains only one solution. In the above example, a branch cut might be a line from the origin out the negative real axis.

Example 7.2.   Find $1^{1/2}$ and represent it on the polar diagram.

$$1^{1/2} = 1\left[\cos\left(\frac{\theta}{2} + k\pi\right) + i\sin\left(\frac{\theta}{2} + k\pi\right)\right]$$

and since $\theta = 0$ in this case

$$1^{1/2} = \cos k\pi + i\sin k\pi$$

There are two distinct roots at $z = +1$ for $k = 0$ and -1 for $k = 1$. The two values are shown in Fig. 7.3. The two solutions are called branches of $\sqrt{1}$, and an appropriate branch cut might be from the origin out the positive imaginary axis, leaving as the single solution 1.

Example 7.3.   Find the roots of $(1 + i)^{1/4}$.

Making use of Eq. (7.13) with $m = 1$ and $n = 4$, $r = \sqrt{2}$, $\theta = \pi/4$, we find that

$$(1 + i)^{1/4} = (\sqrt{2})^{1/4}\left[\cos\left(\frac{\pi}{16} + \frac{2k\pi}{4}\right) + i\sin\left(\frac{\pi}{16} + \frac{2k\pi}{4}\right)\right] \qquad k = 0, 1, 2, 3$$

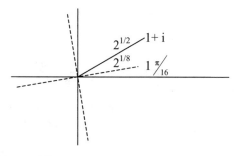

FIGURE 7.4: The roots of $(1 + i)^{1/4}$

Hence, the four roots are as follows:

$$(1 + i)^{1/4} = 2^{1/8}\left[\cos\left(\frac{\pi}{16}\right) + i\sin\left(\frac{\pi}{16}\right)\right]$$

$$= 2^{1/8}\left[\cos\left(\frac{\pi}{16} + \frac{\pi}{2}\right) + i\sin\left(\frac{\pi}{16} + \frac{\pi}{2}\right)\right]$$

$$= 2^{1/8}\left[\cos\left(\frac{\pi}{16} + \pi\right) + i\sin\left(\frac{\pi}{16} + \pi\right)\right]$$

$$= 2^{1/8}\left[\cos\left(\frac{\pi}{16} + \frac{3\pi}{2}\right) + i\sin\left(\frac{\pi}{16} + \frac{3\pi}{2}\right)\right]$$

The locations of the roots are shown in Fig. 7.4.

The natural logarithm can be defined by writing $z = re^{i\theta}$ for $-\pi \leq \theta < \pi$ and noting that

$$\ln z = \ln r + i\theta \tag{7.15}$$

and since z is not affected by adding $2n\pi$ to θ this expression can also be written as

$$\ln z = \ln r + i(\theta + 2n\pi) \quad \text{with} \quad n = 0, 1, 2, \ldots \tag{7.16}$$

When $n = 0$ we obtain the *principal branch*. All of the single valued branches are analytic for $r > 0$ and $\theta_0 < \theta < \theta_0 + 2\pi$.

7.1.1   Limits and Differentiation of Complex Variables: Analytic Functions

Consider a *function of a complex variable* $f(z)$. We generally write

$$f(z) = u(x, y) + iv(x, y)$$

where u and v are real functions of x and y. The derivative of a complex variable is defined as follows:

$$f' = \lim_{\Delta z \to 0} \frac{f(z + \Delta z) - f(z)}{\Delta z} \qquad (7.17)$$

or

$$f'(z) = \lim_{\Delta x, \Delta y \to 0} \frac{u(x + \Delta x, y + \Delta y) + iv(x + \Delta x, y + \Delta y) - u(x, y) - iv(x, y)}{\Delta x + i \Delta y} \qquad (7.18)$$

Taking the limit on Δx first, we find that

$$f'(z) = \lim_{\Delta y \to 0} \frac{u(x, y + \Delta y) + iv(x, y + \Delta y) - u(x, y) - iv(x, y)}{i \Delta y} \qquad (7.19)$$

and now taking the limit on Δy,

$$f'(z) = \frac{1}{i} \frac{\partial u}{\partial y} + \frac{\partial v}{\partial y} = \frac{\partial v}{\partial y} - i \frac{\partial u}{\partial y} \qquad (7.20)$$

Conversely, taking the limit on Δy first,

$$j'(z) = \lim_{\Delta x \to 0} \frac{u(x + \Delta x, y) + iv(x + \Delta x, y) - u(x, y) - iv(x, y)}{\Delta x} \qquad (7.21)$$

$$= \frac{\partial u}{\partial x} + i \frac{\partial v}{\partial x}$$

The derivative exists only if

$$\frac{\partial u}{\partial x} = \frac{\partial v}{\partial y} \quad \text{and} \quad \frac{\partial u}{\partial y} = -\frac{\partial v}{\partial x} \qquad (7.22)$$

These are called the *Cauchy—Riemann conditions*, and in this case the function is said to be *analytic*. If a function is analytic for all x and y it is *entire*.

Polynomials are entire as are trigonometric and hyperbolic functions and exponential functions. We note in passing that analytic functions share the property that both real and imaginary parts satisfy the equation $\nabla^2 u = \nabla^2 v = 0$ in two-dimensional space. It should be obvious at this point that this is important in the solution of the steady-state diffusion equation

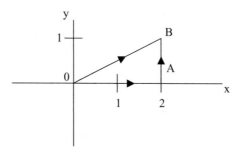

FIGURE 7.5: Integration of an analytic function along two paths

in two dimensions. We mention here that it is also important in the study of incompressible, inviscid fluid mechanics and in other areas of science and engineering. You will undoubtedly meet with it in some of you clurses.

Example 7.4.

$$f = z^2 \quad f' = 2z$$
$$f = \sin z \quad f' = \cos z$$
$$f = e^{az} \quad f' = ae^{az}$$

Integrals
Consider the line integral along a curve C defined as $x = 2y$ from the origin to the point $x = 2, \ y = 1$, path OB in Fig. 7.5.

$$\int_C z^2 dz$$

We can write

$$z^2 = x^2 - y^2 + 2ixy = 3y^2 + 4y^2 i$$

and $dz = (2+i)dy$
Thus

$$\int_{y=0}^{1} (3y^2 + 4y^2 i)(2+i)dy = (3+4i)(2+i)\int_{y=0}^{1} y^2 dy = \frac{2}{3} + \frac{11}{3}i$$

On the other hand, if we perform the same integral along the x axis to $x = 2$ and then along the vertical line $x = 2$ to the same point, path OAB in Fig. 7.5, we find that

$$\int_{x=0}^{2} x^2 dx + \int_{y=0}^{1} (2+iy)^2 i dy = \frac{8}{3} + i \int_{y=0}^{1} (4 - y^2 + 4iy) dy = \frac{2}{3} + \frac{11}{3} i$$

This happened because the function z^2 is *analytic within the region between the two curves*. In general, if a function is analytic in the region contained between the curves, the integral

$$\int_{C} f(z) dz \tag{7.23}$$

is *independent of the path of* C. Since any two integrals are the same, and since if we integrate the first integral along BO only the sign changes, we see that the integral around the closed contour is zero.

$$\oint_{C} f(z) dz = 0 \tag{7.24}$$

This is called the *Cauchy–Goursat theorem* and is true as long as the region R within the closed curve C is *simply connected* and the function is analytic everywhere within the region. A *simply connected region* R is one in which every closed curve within it encloses only points in R.

The theorem can be extended to allow for multiply connected regions. Fig. 7.6 shows a doubly connected region. The method is to make a cut through part of the region and to integrate counterclockwise around C_1, along the path C_2 through the region, clockwise around the interior curve C_3, and back out along C_4. Clearly, the integral along C_2 and C_4 cancels, so that

$$\oint_{C_1} f(z) dz + \oint_{C_3} f(z) dz = 0 \tag{7.25}$$

where the first integral is counterclockwise and second clockwise.

7.1.2   The Cauchy Integral Formula

Now consider the following integral:

$$\oint_{C} \frac{f(z) dz}{(z - z_0)} \tag{7.26}$$

If the function $f(z)$ is analytic then the integrand is also analytic at all points except $z = z_0$. We now form a circle C_2 of radius r_0 around the point $z = z_0$ that is small enough to fit inside

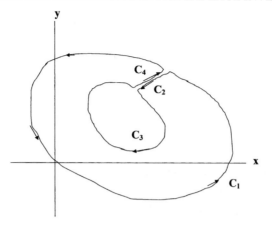

FIGURE 7.6: A doubly connected region

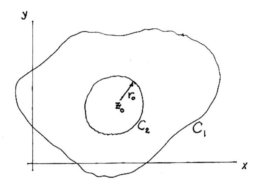

FIGURE 7.7: Derivation of Cauchy's integral formula

the curve C_1 as shown in Fig. 7.7. Thus we can write

$$\oint_{C_1} \frac{f(z)}{z - z_0} dz - \oint_{C_2} \frac{f(z)}{z - z_0} dz = 0 \qquad (7.27)$$

where both integrations are counterclockwise. Let r_0 now approach zero so that in the second integral z approaches z_0, $z - z_0 = r_0 e^{i\theta}$ and $dz = r_0 i e^{i\theta} d\theta$. The second integral is as follows:

$$\oint_{C_2} \frac{f(z_0)}{r_0 e^{i\theta}} r_0 i e^{i\theta} d\theta = -f(z_0) i \int_{\theta=0}^{2\pi} d\theta = -2\pi i f(z_0)$$

Thus, *Cauchy's integral formula* is

$$f(z_0) = \frac{1}{2\pi i} \oint_C \frac{f(z)}{z - z_0} dz \qquad (7.28)$$

where the integral is taken counterclockwise and $f(z)$ is analytic inside C.

We can formally differentiate the above equation n times with respect to z_0 and find an extension as

$$f^{(n)}(z_0) = \frac{n!}{2\pi i} \oint_C \frac{f(z)}{(z-z_0)^{n+1}} dz \qquad (7.29)$$

Problems

1. Show that

   (a) $\sinh z = \sinh x \cos y + i \cosh x \sin y$

   (b) $\cos z = \cos x \cosh y - i \sin x \sinh y$

   and show that each is *entire*.

2. Find all of the values of

   (a) $(-1 + i\sqrt{3})^{\frac{3}{2}}$

   (b) $8^{\frac{1}{6}}$

3. Find all the roots of the equation

   $\sin z = \cosh 4$

4. Find all the zeros of

   (a) $\sinh z$

   (b) $\cosh z$

CHAPTER 8

Solutions with Laplace Transforms

In this chapter, we present detailed solutions of some boundary value problems using the Laplace transform method. Problems in both mechanical vibrations and diffusion are presented along with the details of the inversion method.

8.1 MECHANICAL VIBRATIONS

Example 8.1. Consider an elastic bar with one end of the bar fixed and a constant force F per unit area at the other end acting parallel to the bar. The appropriate partial differential equation and boundary and initial conditions for the displacement $y(x, t)$ are as follows:

$$y_{\tau\tau} = y_{\zeta\zeta}, 0 < \zeta < 1, \quad t > 0$$
$$y(\zeta, 0) = y_t(\zeta, 0) = 0$$
$$y(0, \tau) = 0$$
$$y_\zeta(1, \tau) = F/E = g$$

We obtain the Laplace transform of the equation and boundary conditions as

$$s^2 Y = Y_{\zeta\zeta}$$
$$Y(s, 0) = 0$$
$$Y_\zeta(s, 1) = g/s$$

Solving the differential equation for $Y(s, \zeta)$,

$$Y(s) = (A \sinh \varsigma s + B \cosh \varsigma s)$$

Applying the boundary conditions we find that $B = 0$ and

$$\frac{g}{s} = As \cosh s$$
$$A = \frac{g}{s^2 \cosh s}$$
$$Y(s) = \frac{g \sinh \varsigma s}{s^2 \cosh s}$$

Since the function $\frac{1}{s}\sinh\varsigma s = \varsigma + \frac{s^2\varsigma^3}{3!} + \frac{s^4\varsigma^5}{5!} + \ldots$ the function

$$\frac{1}{s}\sinh\varsigma s$$

is analytic and $Y(s)$ can be written as the ratio of two analytic functions

$$Y(s) = \frac{\frac{1}{s}\sinh\varsigma s}{s\cosh s}$$

$Y(s)$ therefore has a simple pole at $s = 0$ and the residue there is

$$R(s = 0) = \frac{\lim}{s \to 0} s Y(s) = \frac{\lim}{s \to 0} \frac{\varsigma + \frac{s^2\varsigma^3}{3!} + \cdots}{\cosh s} = g\varsigma$$

The remaining poles are the singularities of $\cosh s$. But $\cosh s = \cosh x \cos y + i \sinh x \sin y$, so the zeros of this function are at $x = 0$ and $\cos y = 0$.

Hence, $s_n = i(2n - 1)\pi/2$. The residues at these points are

$$R(s = s_n) = \frac{\lim}{s \to s_n}\left[\frac{g\sinh\varsigma s}{s\frac{d}{ds}(s\cosh s)}\right]e^{st} = \frac{g}{s_n^2}\frac{\sinh\varsigma s_n}{\sinh s_n}e^{s_n t}\,(n = \pm1, \pm2, \pm3\ldots)$$

Since

$$\sinh\left[i\frac{2n-1}{2}(\pi\varsigma)\right] = i\sin\left[\frac{2n-1}{2}(\pi\varsigma)\right]$$

we have

$$R(s = s_n) = \frac{gi\sin\left[\frac{2n-1}{2}(\pi\varsigma)\right]}{-\left[\frac{2n-1}{2}\pi\right]^2 i\sin\left[\frac{2n-1}{2}\pi\right]}\exp\left[i\frac{2n-1}{2}\pi\tau\right]$$

and

$$\sin\left[\frac{2n-1}{s}\pi\right] = (-1)^{n+1}$$

The exponential function can be written as

$$\exp\left[i\frac{2n-1}{2}\pi\tau\right] = \cos\left[\frac{2n-1}{2}\pi\tau\right] + i\sin\left[\frac{2n-1}{2}\pi\tau\right]$$

Note that for the poles on the negative imaginary axis $(n < 0)$ this expression can be written as

$$\exp\left[i\frac{2m-1}{2}\pi\tau\right] = \cos\left[\frac{2m-1}{2}\pi\tau\right] - i\sin\left[\frac{2m-1}{2}\pi\tau\right]$$

where $m = -n > 0$. This corresponds to the conjugate poles.

Thus for *each* of the sets of poles we have

$$R(s = s_n) = \frac{4g(-1)^n}{\pi^2(2n-1)^2} \sin \frac{(2n-1)\pi\varsigma}{2} \exp\left[\frac{(2n-1)\pi\tau i}{2}\right]$$

Now adding the residues corresponding to each pole and its conjugate we find that the final solution is as follows:

$$y(\varsigma, \tau) = g\left[\varsigma + \frac{8}{\pi^2} \sum_{n=1}^{\infty} \frac{(-1)^n}{(2n-1)^2} \sin \frac{(2n-1)\pi\varsigma}{2} \cos \frac{(2n-1)\pi\tau}{2}\right]$$

Suppose that instead of a constant force at $\varsigma = 1$, we allow g to be a function of τ. In this case, the Laplace transform of $y(\varsigma, \tau)$ takes the form

$$Y(\varsigma, s) = \frac{G(s) \sinh(\varsigma s)}{s \cosh s}$$

The simple pole with residue $g\varsigma$ is not present. However, the other poles are still at the same s_n values. The residues at each of the conjugate poles of the function

$$F(s) = \frac{\sinh(\varsigma s)}{s \cosh s}$$

are

$$\frac{2(-1)^n}{\pi(2n-1)} \sin \frac{(2n-1)\pi\varsigma}{2} \sin \frac{(2n-1)\pi\tau}{2} = f(\varsigma, \tau)$$

According to the convolution theorem

$$y(\varsigma, \tau) = \int_{\tau'=0}^{\tau} y(\tau - \tau')g(\tau')d\tau'$$

$$y(\varsigma, \tau) = \frac{4}{\pi} \sum_{n=0}^{\infty} \frac{(-1)^n}{(2n-1)} \sin \frac{(2n-1)\pi\varsigma}{2} \int_{\tau'}^{\tau} g(\tau - \tau') \sin \frac{(2n-1)\pi\tau'}{2} d\tau'.$$

In the case that $g = $ constant, integration recovers the previous equation.

Example 8.2.   An infinitely long string is initially at rest when the end at $x = 0$ undergoes a transverse displacement $y(0, t) = f(t)$. The displacement is described by the differential equation and boundary conditions as follows:

$$\frac{\partial^2 y}{\partial t^2} = \frac{\partial^2 y}{\partial x^2}$$

$$y(x, 0) = y_t(x, 0) = 0$$

$$y(0, t) = f(t)$$

$$y \text{ is bounded}$$

Taking the Laplace transform with respect to time and applying the initial conditions yields

$$s^2 Y(x, s) = \frac{d^2 Y(x, s)}{dx^2}$$

The solution may be written in terms of exponential functions

$$Y(x, s) = A e^{-sx} + B e^{sx}$$

In order for the solution to be bounded $B = 0$. Applying the condition at $x = 0$ we find

$$A = F(s)$$

where $F(s)$ is the Laplace transform of $f(t)$.

Writing the solution in the form

$$Y(x, s) = s F(s) \frac{e^{-sx}}{s}$$

and noting that the inverse transform of e^{-sx}/s is the Heaviside step $U_x(t)$ where

$$U_x(t) = 0 \quad t < x$$

$$U_x(t) = 1 \quad t > x$$

and that the inverse transform of $s F(s)$ is $f'(t)$, we find using convolution that

$$y(x, t) = \int_{\mu=0}^{t} f'(t - \mu) U_x(\mu) d\mu = f(t - x) \quad x < t$$

$$= 0 \quad x > t$$

For example, if $f(t) = \sin \omega t$

$$y(x, t) = \sin \omega (t - x) \quad x < t$$

$$= 0 \quad x > t$$

Problems

1. Solve the above vibration problem when

$$y(0, \tau) = 0$$

$$y(1, \tau) = g(\tau)$$

Hint: To make use of convolution see Example 8.3.

2. Solve the problem

$$\frac{\partial^2 y}{\partial t^2} = \frac{\partial^2 y}{\partial x^2}$$
$$y_x(0, t) = y(x, 0) = y_t(x, 0) = 0$$
$$y(1, t) = h$$

using the Laplace transform method.

8.2   DIFFUSION OR CONDUCTION PROBLEMS

We now consider the conduction problem

Example 8.3.

$$u_\tau = u_{\varsigma\varsigma}$$
$$u(1, \tau) = f(\tau)$$
$$u(0, \tau) = 0$$
$$u(\varsigma, 0) = 0$$

Taking the Laplace transform of the equation and boundary conditions and noting that $u(\varsigma, 0) = 0$,

$$s U(s) = U_{\varsigma\varsigma}$$

solution yields

$$U = A \sinh \sqrt{s}\,\varsigma + B \cosh \sqrt{s}\,\varsigma$$

$$U(0, s) = 0$$

$$U(1, s) = F(s)$$

The first condition implies that $B = 0$ and the second gives

$$F(s) = A \sinh \sqrt{s}$$

and so $U = F(s) \frac{\sinh \sqrt{s}\,\varsigma}{\sinh \sqrt{s}}$.

If $f(\tau) = 1$, $F(s) = 1/s$, a particular solution, V, is

$$V = \frac{\sinh \sqrt{s}\,\varsigma}{s \, \sinh \sqrt{s}}$$

where

$$v = L^{-1}V(s)$$

Now,

$$\frac{\sinh \sqrt{s}\,\varsigma}{\sinh \sqrt{s}} = \frac{\varsigma\sqrt{s} + \frac{(\varsigma\sqrt{s})^3}{3!} + \frac{(\varsigma\sqrt{s})^5}{5!} + \cdots}{\sqrt{s} + \frac{(\sqrt{s})^3}{3!} + \frac{(\sqrt{s})^5}{5!} + \cdots}$$

and so there is a simple pole of $Ve^{s\tau}$ at $s = 0$. Also, since when $\sinh\sqrt{s} = 0$, $\sinh\varsigma\sqrt{s}$ not necessarily zero, there are simple poles at $\sinh\sqrt{s} = 0$ or $s = -n^2\pi^2$. The residue at the pole $s = 0$ is

$$\lim_{s \to 0} s\, V(s)e^{s\tau} = \varsigma$$

and since $V(s)\,e^{s\tau}$ has the form $P(s)/Q(s)$ the residue of the pole at $-n^2\pi^2$ is

$$\frac{P(\varsigma, -n^2\pi^2)}{Q'(-n^2\pi^2)}e^{-n^2\pi^2\tau} = \left. \frac{\sinh \varsigma\sqrt{s}e^{-n^2\pi^2\tau}}{\frac{\sqrt{s}}{2}\cosh\sqrt{s} + \sinh\sqrt{s}} \right]_{s=-n^2\pi^2} = 2\frac{\sin(n\pi\varsigma)}{n\pi\cos(n\pi)}e^{-n^2\pi^2\tau}$$

The solution for $v(\varsigma, \tau)$ is then

$$v(\varsigma, \tau) = \varsigma + \sum_{n=1}^{\infty} \frac{2(-1)^n}{n\pi}e^{-n^2\pi^2\tau}\sin(n\pi\varsigma)$$

The solution for the general case as originally stated with $u(1, \tau) = f(\tau)$ is obtained by first differentiating the equation for $v(\varsigma, \tau)$ and then noting the following:

$$U(\varsigma, s) = s\,F(s)\frac{\sinh \varsigma\sqrt{s}}{s \, \sinh \sqrt{s}}$$

and

$$L\left[f'(\tau)\right] = s\,F(s) - f(\tau = 0)$$

so that

$$U(\varsigma, s) = f(\tau = 0)V(\varsigma, s) + L\left[f'(s)\right]V(\varsigma, s)$$

Consequently

$$u(\varsigma, \tau) = f(\tau = 0)v(\varsigma, \tau) + \int_{\tau'=0}^{\tau} f'(\tau - \tau')v(\varsigma, \tau')d\tau'$$

$$= \varsigma f(\tau) + \frac{2f(0)}{\pi} \sum_{n=1}^{\infty} \frac{(-1)^n}{n} e^{-n^2\pi^2\tau} \sin(n\pi\varsigma)$$

$$+ \frac{2}{\pi} \sum_{n=1}^{\infty} \frac{(-1)^n}{n} \sin(n\pi\varsigma) \int_{\tau'=0}^{\tau} f'(\tau - \tau')e^{-n^2\pi^2\tau'}d\tau'$$

This series converges rapidly for large values of τ. However for small values of τ, it converges slowly. *There is another form of solution that converges rapidly for small τ.*

The Laplace transform of $v(\zeta, \tau)$ can be written as

$$\frac{\sinh \varsigma\sqrt{s}}{s \sinh \sqrt{s}} = \frac{e^{\varsigma\sqrt{s}} - e^{-\varsigma\sqrt{s}}}{s(e^{\sqrt{s}} - e^{-\sqrt{s}})} = \frac{1}{s e^{\sqrt{s}}} \frac{e^{\varsigma\sqrt{s}} - e^{-\varsigma\sqrt{s}}}{1 - e^{-2\sqrt{s}}}$$

$$= \frac{1}{s e^{\sqrt{s}}} \left[e^{\varsigma\sqrt{s}} - e^{-\varsigma\sqrt{s}} \right] \left[1 + e^{-2\sqrt{s}} + e^{-4\sqrt{s}} + e^{-6\sqrt{s}} + \cdots \right]$$

$$= \frac{1}{s} \sum_{n=0}^{\infty} \left[e^{-(1+2n-\varsigma)\sqrt{s}} - e^{-(1+2n+\varsigma)\sqrt{s}} \right]$$

The inverse Laplace transform of $\frac{e^{-k\sqrt{s}}}{s}$ is the complimentary error function, defined by

$$\text{erfc}(k/2\sqrt{\tau}) = 1 - \frac{2}{\sqrt{\pi}} \int_{x=0}^{k/2\sqrt{\tau}} e^{-x^2} dx$$

Thus we have

$$v(\varsigma, \tau) = \sum_{n=0}^{\infty} \left[\text{erfc}\left(\frac{1+2n-\varsigma}{2\sqrt{\tau}} \right) - \text{erfc}\left(\frac{1+2n+\varsigma}{2\sqrt{\tau}} \right) \right]$$

and this series converges rapidly for small values of τ.

Example 8.4.   Next we consider a conduction problem with a convective boundary condition:

$$u_\tau = u_{\varsigma\varsigma}$$
$$u(\tau, 0) = 0$$
$$u_\varsigma(\tau, 1) + Hu(\tau, 1) = 0$$
$$u(0, \varsigma) = \varsigma$$

Taking the Laplace transform

$$sU - \varsigma = U_{\varsigma\varsigma}$$

$$U(s, 0) = 0$$

$$U_\varsigma(s, 1) + HU(s, 1) = 0$$

The differential equation has a homogeneous solution

$$U_h = A\cosh(\sqrt{s}\,\varsigma) + B\sinh(\sqrt{s}\,\varsigma)$$

and a particular solution

$$U_p = \frac{\varsigma}{s}$$

so that

$$U = \frac{\varsigma}{s} + A\cosh(\sqrt{s}\,\varsigma) + B\sinh(\sqrt{s}\,\varsigma)$$

Applying the boundary conditions, we find $A = 0$

$$B = -\frac{1 + H}{s\left[\sqrt{s}\cosh(\sqrt{s}) + H\sinh(\sqrt{s})\right]}$$

The Laplace transform of the solution is as follows:

$$U = \frac{\varsigma}{s} - \frac{(1 + H)\sinh(\sqrt{s}\,\varsigma)}{s\left[\sqrt{s}\cosh(\sqrt{s}) + H\sinh(\sqrt{s})\right]}$$

The inverse transform of the first term is simply ς. For the second term, we must first find the poles. There is an isolated pole at $s = 0$. To obtain the residue of this pole note that

$$\frac{\lim}{s \to 0} - \frac{(1 + H)\sinh\varsigma\sqrt{s}}{\sqrt{s}\cosh\sqrt{s} + H\sinh\sqrt{s}}e^{st} = \frac{\lim}{s \to 0} - \frac{(1 + H)(\varsigma\sqrt{s} + \cdots)}{\sqrt{s} + H(\sqrt{s} + \cdots)} = -\varsigma$$

canceling the first residue. To find the remaining residues let $\sqrt{s} = x + iy$. Then

$$(x + iy)[\cosh x \cos y + i\sinh x \sin y] + H[\sinh x \cos y + i\cosh x \sin y] = 0$$

Setting real and imaginary parts equal to 0 yields

$$x\cosh x \cos y - y\sinh x \sin y + H\sinh x \cos y = 0$$

and

$$y\cosh x \cos y + x\sinh x \sin y + H\cosh x \sin y = 0$$

which yields

$$x = 0$$

$$y \cos y + H \sin y = 0$$

The solution for the second term of U is

$$\lim_{s \to iy} \frac{(s - iy)(1 + H) \sinh(\sqrt{s}\varsigma)e^{s\tau}}{s \left[\sqrt{s} \cosh(\sqrt{s}) + H \sinh \sqrt{s})\right]}$$

or

$$\left[\frac{P(\varsigma, s)e^{s\tau}}{Q'(\varsigma, s)}\right]_{s=-y^2}$$

where

$$Q = s \left[\sqrt{s} \cosh \sqrt{s} + H \sinh \sqrt{s}\right]$$

$$Q' = \sqrt{s} \cosh \sqrt{s} + H \sinh \sqrt{s} + s \left[\frac{1}{2\sqrt{s}} \cosh \sqrt{s} + \frac{1}{2} \sinh \sqrt{s} + \frac{H}{2\sqrt{s}} \cosh \sqrt{s}\right]$$

$$Q' = \frac{\sqrt{s}(1 + H)}{2} \cosh \sqrt{s} + \frac{s}{2} \sinh \sqrt{s}$$

$$Q' = \left[\frac{\sqrt{s}(1 + H)}{2} - \frac{s\sqrt{s}}{2H}\right] \cosh \sqrt{s}$$

$$Q'(s = -y^2) = \left[\frac{H(H + 1) + y^2}{2H}\right] iy \cos(y)$$

while

$$P(s = -y^2) = (1 + H)i \sin(y\varsigma)e^{-y^2\tau}$$

$$u_n(\varsigma, \tau) = \frac{-(1 + H) \sin(y_n\varsigma)e^{-y^2\tau}}{\left[\frac{H(H+1)+y^2}{2H}\right] y_n \cos(y_n)} = \frac{-2H(H + 1) \sin(y_n\varsigma)e^{-y^2\tau}}{[H(H + 1) + y^2] y_n \cos(y_n)}$$

$$= \left[\frac{2(H + 1)}{H(H + 1) + y^2}\right] \frac{\sin \varsigma y_n}{\sin y_n} e^{-y_n^2\tau}$$

The solution is therefore

$$u(\varsigma, \tau) = \sum_{n=1}^{\infty} \frac{2(H + 1)}{H(H + 1) + y^2} \frac{\sin \varsigma y_n}{\sin y_n} e^{-y_n^2\tau}$$

Note that as a partial check on this solution, we can evaluate the result when $H \to \infty$ as

$$u(\varsigma, \tau) = \sum_{n=1}^{\infty} \frac{-2}{y_n \cos y_n} \sin \varsigma \, y_n e^{-y_n^2 \tau} = \sum_{n=1}^{\infty} \frac{2(-1)^{n+1}}{n\pi} \sin(n\pi \varsigma) e^{-n^2 \pi^2 \tau}$$

in agreement with the separation of variables solution. Also, letting $H \to 0$ we find

$$u(\varsigma, \tau) = \sum_{n=1}^{\infty} \frac{2}{y_n^2} \frac{\sin(y_n \varsigma)}{\sin(y_n)} e^{y_n^2 \tau}$$

with $y_n = \frac{2n-1}{2}\pi$ again in agreement with the separation of variables solution.

Example 8.5. Next we consider a conduction (diffusion) problem with a transient source $q(\tau)$. (Nondimensionalization and normalization are left as an exercise.)

$$u_\tau = u_{\varsigma\varsigma} + q(\tau)$$
$$u(\varsigma, 0) = 0 = u_\varsigma(0, \tau)$$
$$u(1, \tau) = 1$$

Obtaining the Laplace transform of the equation and boundary conditions we find

$$s U = U_{\varsigma\varsigma} + Q(s)$$
$$U_\varsigma(0, s) = 0$$
$$U(1, s) = \frac{1}{s}$$

A particular solution is

$$U_P = \frac{Q(s)}{s}$$

and the homogeneous solution is

$$U_H = A \sinh(\varsigma \sqrt{s}) + B \cosh(\varsigma \sqrt{s})$$

Hence the general solution is

$$U = \frac{Q}{s} + A \sinh(\varsigma \sqrt{s}) + B \cosh(\varsigma, \sqrt{s})$$

Using the boundary conditions

$$U_\varsigma(0, s) = 0, \quad A = 0$$

$$U(1, s) = \frac{1}{s} = \frac{Q}{s} + B\cosh(\sqrt{s}) \quad B = \frac{1 - Q}{s\,\cosh(\sqrt{s})}$$

$$U = \frac{Q}{s} + \frac{1 - Q}{s}\frac{\cosh(\varsigma\sqrt{s})}{\cosh(\sqrt{s})}$$

The poles are (with $\sqrt{s} = x + iy$)

$$\cosh\sqrt{s} = 0 \quad \text{or} \quad \cos y = 0 \quad \sqrt{s} = \pm\frac{2n - 1}{2}\pi\,i$$

$$s = -\left(\frac{2n - 1}{2}\right)^2 \pi^2 = -\lambda_n^2 \quad n = 1, 2, 3, \ldots$$

or when $s = 0$.

When $s = 0$ the residue is

$$\text{Res} = \frac{\lim}{s \to 0} s\,U(s)e^{st} = 1$$

The denominator of the second term is $s \cosh\sqrt{s}$ and its derivative with respect to s is

$$\cosh\sqrt{s} + \frac{\sqrt{s}}{2}\sinh\sqrt{s}$$

When $s = -\lambda_n^2$, we have for the residue of the second term

$$\frac{\lim}{s \to -\lambda_n^2}\left[\frac{(1 - Q)\cosh(\varsigma\sqrt{s})}{\cosh\sqrt{s} + \frac{\sqrt{s}}{2}\sinh\sqrt{s}}\right]e^{st}$$

and since

$$\sinh\sqrt{s} = i\sin\left(\frac{2n - 1}{2}\right)\pi = i(-1)^{n+1}$$

and

$$\cosh(\varsigma\sqrt{s}) = \cos\left(\frac{2n - 1}{2}\right)\varsigma\pi$$

we have

$$L^{-1}\frac{\cosh(\varsigma\sqrt{s})}{s\cosh\sqrt{s}} = \frac{\cos\left(\frac{2n-1}{2}\varsigma\pi\right)}{\left(\frac{2n-1}{2}\right)\pi\,i^2(-1)^{n+1}}e^{-\left(\frac{2n-1}{2}\right)^2\pi^2\tau} = \frac{2(-1)^n\cos\left(\frac{2n-1}{2}\varsigma\pi\right)}{(2n - 1)\pi}e^{-\left(\frac{2n-1}{2}\right)^2\pi^2\tau}$$

We now use the convolution principle to evaluate the solution for the general case of $q(\tau)$. We are searching for the inverse transform of

$$\frac{1}{s}\frac{\cosh(\varsigma\sqrt{s})}{\cosh\sqrt{s}} + \frac{Q(s)}{s}\left(1 - \frac{\cosh(\varsigma\sqrt{s})}{\cosh\sqrt{s}}\right)$$

The inverse transform of the first term is given above. As for the second term, the inverse transform of $Q(s)$ is simply $q(\tau)$ and the inverse transform of the second term, absent $Q(s)$ is

$$1 - \frac{2(-1)^{n+1}\cos\left(\frac{2n-1}{2}\varsigma\pi\right)}{(2n-1)\pi}e^{-\left(\frac{2n-1}{2}\right)^2\pi^2\tau}$$

According to the convolution principle, and summing over all poles

$$u(\varsigma,\tau) = \sum_{n=1}^{\infty}\frac{2(-1)^{n+1}\cos\left(\frac{2n-1}{2}\varsigma\pi\right)}{(2n-1)\pi}e^{-\left(\frac{2n-1}{2}\right)^2\pi^2\tau}$$

$$+ \sum_{n=1}^{\infty}\int_{\tau'=0}^{\tau}\left[1 - \frac{2(-1)^{n+1}\cos\left(\frac{2n-1}{2}\varsigma\pi\right)}{(2n-1)\pi}\right]e^{-\left(\frac{2n-1}{2}\right)^2\pi^2\tau}q(\tau-\tau')d\tau'$$

Example 8.6. Next consider heat conduction in a semiinfinite region $x > 0$, $t > 0$. The initial temperature is zero and the wall is subjected to a temperature $u(0, t) = f(t)$ at the $x = 0$ surface.

$$u_t = u_{xx}$$
$$u(x, 0) = 0$$
$$u(0, t) = f(t)$$

and u is bounded.

Taking the Laplace transform and applying the initial condition

$$sU = U_{xx}$$

Thus

$$U(x, s) = A\sinh x\sqrt{s} + B\cosh x\sqrt{s}$$

Both functions are unbounded for $x \to \infty$. Thus it is more convenient to use the equivalent solution

$$U(x, s) = Ae^{-x\sqrt{s}} + Be^{x\sqrt{s}} = Ae^{-x\sqrt{s}}$$

in order for the function to be bounded. Applying the boundary condition at $x = 0$

$$F(s) = A$$

Thus we have

$$U(x, s) = F(s)e^{-x\sqrt{s}}$$

Multiplying and dividing by s gives

$$U(x, s) = s F(s)\frac{e^{-x\sqrt{s}}}{s}$$

The inverse transform of $e^{-x\sqrt{s}}/s$ is

$$L^{-1}\left[\frac{e^{-x\sqrt{s}}}{s}\right] = \text{erfc}\left(\frac{x}{2\sqrt{t}}\right)$$

and we have seen that

$$L\{f'\} = s F(s) - f(0)$$

Thus, making use of convolution, we find

$$u(x, t) = f(0)\text{erfc}\left(\frac{x}{2\sqrt{t}}\right) + \int_{\mu=0}^{t} f'(t - \mu)\,\text{erfc}\frac{x}{2\sqrt{\mu}}d\mu$$

Example 8.7. Now consider a problem in cylindrical coordinates. An infinite cylinder is initially at dimensionless temperature $u(r, 0) = 1$ and dimensionless temperature at the surface $u(1, t) = 0$. We have

$$\frac{\partial u}{\partial t} = \frac{1}{r}\frac{\partial}{\partial r}\left(r\frac{\partial u}{\partial r}\right)$$

$$u(1, t) = 0$$

$$u(r, 0) = 1$$

$$u \text{ bounded}$$

The Laplace transform with respect to time yields

$$s U(r, s) - 1 = \frac{1}{r}\frac{d}{dr}\left(r\frac{dU}{dr}\right)$$

with

$$U(1, s) = \frac{1}{s}$$

Obtaining the homogeneous and particular solutions yields

$$U(r, s) = \frac{1}{s} + AJ_0(i\sqrt{s}r) + BY_0(i\sqrt{s}r)$$

The boundedness condition requires that $B = 0$, while the condition at $r = 1$

$$A = -\frac{1}{sJ_0(i\sqrt{s})}$$

Thus

$$U(r, s) = \frac{1}{s} - \frac{J_0(i\sqrt{s}r)}{sJ_0(i\sqrt{s})}$$

The inverse transform is as follows:

$$u(r, t) = 1 - \sum \text{Residues of} \left[e^{st} \frac{J_0(i\sqrt{s}r}{sJ_0(i\sqrt{s})} \right]$$

Poles of the function occur at $s = 0$ and $J_0(i\sqrt{s}) = 0$ or $i\sqrt{s} = \lambda_n$, the roots of the Bessel function of the first kind order are zero. Thus, they occur at $s = -\lambda_n^2$. The residues are

$$\lim_{s \to 0} \left[e^{st} \frac{J_0(i\sqrt{s}r)}{J_0(i\sqrt{s})} \right] = 1$$

and

$$\lim_{s \to -\lambda_n^2} \left[e^{st} \frac{J_0(i\sqrt{s}r)}{sJ_0'(i\sqrt{s})} \right] = \lim_{s \to -\lambda_n^2} \left[e^{st} \frac{J_0(i\sqrt{s}r)}{-J_1(i\sqrt{s})i/2\sqrt{s}} \right] = e^{-\lambda_n^2 t} \left[\frac{J_0(\lambda_n r)}{-\frac{1}{2}\lambda_n J_1(\lambda_n)} \right]$$

The two unity residues cancel and the final solution is as follows:

$$u(r, t) = \sum_{n=1}^{\infty} e^{-\lambda_n^2 t} \frac{J_0(\lambda_n r)}{\lambda_n J_1(\lambda_n)}$$

Problems

1. Consider a finite wall with initial temperature zero and the wall at $x = 0$ insulated. The wall at $x = 1$ is subjected to a temperature $u(1, t) = f(t)$ for $t > 0$. Find $u(x, t)$.

2. Consider a finite wall with initial temperature zero and with the temperature at $x = 0$ $u(0, t) = 0$. The temperature gradient at $x = 1$ suddenly becomes $u_x(1, t) = f(t)$ for $t > 0$. Find the temperature when $f(t) = 1$ and for general $f(t)$.

3. A cylinder is initially at temperature $u = 1$ and the surface is subject to a convective boundary condition $u_r(t, 1) + Hu(t, 1) = 0$. Find $u(t, r)$.

8.3   DUHAMEL'S THEOREM

We are now prepared to solve the more general problem

$$\nabla^2 u + g(r, t) = \frac{\partial u}{\partial t} \tag{8.1}$$

where r may be considered a vector, that is, the problem is in three dimensions. The general boundary conditions are

$$\frac{\partial u}{\partial n_i} + h_i u = f_i(r, t) \text{ on the boundary } S_i \tag{8.2}$$

and

$$u(r, 0) = F(r) \tag{8.3}$$

initially. Here $\frac{\partial u}{\partial n_i}$ represents the normal derivative of u at the surface. We present Duhamel's theorem without proof.

Consider the auxiliary problem

$$\nabla^2 P + g(r, \lambda) = \frac{\partial P}{\partial t} \tag{8.4}$$

where λ is a timelike constant with boundary conditions

$$\frac{\partial P}{\partial n_i} + h_i P = f_i(r, \lambda) \text{ on the boundary } S_i \tag{8.5}$$

and initial condition

$$P(r, 0) = F(r) \tag{8.6}$$

The solution of Eqs. (8.1), (8.2), and (8.3) is as follows:

$$u(x, y, z, t) = \frac{\partial}{\partial t} \int_{\lambda=0}^{t} P(x, y, z, \lambda, t - \lambda)d\lambda = F(x, y, z) + \int_{\lambda=0}^{t} \frac{\partial}{\partial t} P(x, y, z, \lambda, t - \lambda)d\lambda \tag{8.7}$$

This is Duhamel's theorem. For a proof, refer to the book by Arpaci.

Example 8.8.   Consider now the following problem with a time-dependent heat source:

$$u_t = u_{xx} + xe^{-t}$$
$$u(0, t) = u(1, t) = 0$$
$$u(x, 0) = 0$$

We first solve the problem

$$P_t = P_{xx} + xe^{-\lambda}$$

$$P(0, t) = P(1, t) = 0$$

$$P(x, 0) = 0$$

while holding λ constant.

Recall from Chapter 2 that one technique in this case is to assume a solution of the form

$$P(x, \lambda, t) = X(x) + W(x, \lambda, t)$$

so that

$$W_t = W_{xx}$$

$$W(0, \lambda, t) = W(1, \lambda, t) = 0$$

$$W(x, \lambda, 0) = -X(x, \lambda)$$

and

$$X_{xx} + xe^{-\lambda} = 0$$

$$X(0) = X(1) = 0$$

Separating variables in the equation for $W(x, t)$, we find that for $W(x, \lambda, t) = S(x)Q(t)$

$$\frac{Q_t}{Q} = \frac{S_{xx}}{S} = -\beta^2$$

The minus sign has been chosen so that Q remains bounded. The boundary conditions on $S(x)$ are as follows:

$$S(0) = S(1) = 0$$

The solution gives

$$S = A\sin(\beta x) + B\cos(\beta x)$$

$$Q = Ce^{-\beta t}$$

Applying the boundary condition at $x = 0$ requires that $B = 0$ and applying the boundary condition at $x = 1$ requires that $\sin(\beta) = 0$ or $\beta = n\pi$.

Solving for $X(x)$ and applying the boundary conditions gives

$$X = \frac{x}{6}(1 - x^2)e^{-\lambda} = -W(x, \lambda, 0)$$

The solution for $W(x, t)$ is then obtained by superposition:

$$W(x, t) = \sum_{n=0}^{\infty} K_n e^{-n^2 \pi^2 t} \sin(n\pi x)$$

and using the orthogonality principle

$$e^{-\lambda} \int_{x=0}^{1} \frac{x}{6}(x^2 - 1) \sin(n\pi x) dx = K_n \int_{n=0}^{1} \sin^2(n\pi x) dx = \frac{1}{2} K_n$$

so

$$W(x, t) = \sum_{n=1}^{\infty} e^{-\lambda} \int_{x=0}^{1} \frac{x}{3}(x^2 - 1) \sin(n\pi x) dx \, e^{-n^2 \pi^2 t} \sin(n\pi x)$$

$$P(x, \lambda, t) = \left\{ \frac{x}{6}(1 - x^2) + \sum_{n=1}^{\infty} \int_{x=0}^{1} \frac{x}{3}(x^2 - 1) \sin(n\pi x) dx \, \sin(n\pi x) \, e^{-n^2 \pi^2 t} \right\} e^{-\lambda}$$

and

$$P(x, \lambda, t - \lambda) = \left\{ \frac{x}{6}(1 - x^2) e^{-\lambda} \right\}$$

$$+ \sum_{n=1}^{\infty} \int_{x=0}^{1} \frac{x}{3}(x^2 - 1) \sin(n\pi x) dx \, \sin(n\pi x) \, e^{-n^2 \pi^2 t} \, e^{n^2 \pi^2 \lambda - \lambda}$$

$$\frac{\partial}{\partial t} P(x, \lambda, t - \lambda) = \sum_{n=1}^{\infty} n^2 \pi^2 \int_{x=0}^{1} \frac{x}{3}(1 - x^2) \sin(n\pi x) dx e^{-n^2 \pi^2 t} \, e^{(n^2 \pi^2 - 1)\lambda}$$

According to Duhamel's theorem, the solution for $u(x, t)$ is then

$$u(x, t) = \sum_{n=1}^{\infty} \int_{x=0}^{1} \frac{x}{3}(1 - x^2) n^2 \pi^2 \sin(n\pi x) dx \, \sin(n\pi x) \int_{\lambda=0}^{t} e^{-n^2 \pi^2 (t-\lambda) - \lambda} d\lambda$$

$$= \sum_{n=1}^{\infty} \frac{n^2 \pi^2}{n^2 \pi^2 - 1} \int_{x=0}^{1} \frac{x}{3}(1 - x^2) \sin(n\pi x) dx \, [e^{-t} - e^{-n^2 \pi^2 t}] \sin(n\pi x)$$

Example 8.9. Reconsider Example 8.6 in which $u_t = u_{xx}$ on the half space, with

$$u(x, 0) = 0$$

$$u(0, t) = f(t)$$

To solve this using Duhamel's theorem, we first set $f(t) = f(\lambda)$ with λ a timelike constant.

Following the procedure outlined at the beginning of Example 8.6, we find

$$U(x, s) = f(\lambda)\frac{e^{-x\sqrt{s}}}{s}$$

The inverse transform is as follows:

$$u(x, t, \lambda) = f(\lambda)\,\text{erfc}\left(\frac{x}{2\sqrt{t}}\right)$$

Using Duhamel's theorem,

$$u(x, t) = \int_{\lambda=0}^{t} \frac{\partial}{\partial t}\left[f(\lambda)\text{erfc}\left(\frac{x}{2\sqrt{t-\lambda}}\right)\right] d\lambda$$

which is a different form of the solution given in Example 8.6.

Problems

1. Show that the solutions given in Examples 8.6 and 8.9 are equivalent.

2. Use Duhamel's theorem along with Laplace transforms to solve the following conduction problem on the half space:

$$u_t = u_{xx}$$
$$u(x, 0) = 0$$
$$u_x(0, t) = f(t)$$

3. Solve the following problem first using separation of variables:

$$\frac{\partial u}{\partial t} = \frac{\partial^2 u}{\partial x^2} + \sin(\pi x)$$
$$u(t, 0) = 0$$
$$u(t, 1) = 0$$
$$u(0, x) = 0$$

4. Consider now the problem

$$\frac{\partial u}{\partial t} = \frac{\partial^2 u}{\partial x^2} + \sin(\pi x)te^{-t}$$

with the same boundary conditions as Problem 7. Solve using Duhamel's theorem.

FURTHER READING

V. S. Arpaci, *Conduction Heat Transfer*, Reading, MA: Addison-Wesley, 1966.

R. V. Churchill, *Operational Mathematics*, 3rd ed. New York: McGraw-Hill, 1972.

I. H. Sneddon, *The Use of Integral Transforms*, New York: McGraw-Hill, 1972.

CHAPTER 9

Sturm–Liouville Transforms

Sturm–Liouville transforms include a variety of examples of choices of the kernel function $K(s, t)$ that was presented in the general transform equation at the beginning of Chapter 6. We first illustrate the idea with a simple example of the Fourier sine transform, which is a special case of a Sturm–Liouville transform. We then move on to the general case and work out some examples.

9.1 A PRELIMINARY EXAMPLE: FOURIER SINE TRANSFORM

Example 9.1. Consider the boundary value problem

$$u_t = u_{xx} \qquad x \leq 0 \leq 1$$

with boundary conditions

$$u(0, t) = 0$$

$$u_x(1, t) + Hu(1, t) = 0$$

and initial condition

$$u(x, 0) = 1$$

Multiply both sides of the differential equation by $\sin(\lambda x)dx$ and integrate over the interval $x \leq 0 \leq 1$.

$$\int_{x=0}^{1} \sin(\lambda x)\frac{d^2 u}{dx^2}\,dx = \frac{d}{dt}\int_{x=0}^{1} u(x, t)\sin(\lambda x)dx$$

Integration of the left hand side by parts yields

$$\int_{x=0}^{1} \frac{d^2}{dx^2}[\sin(\lambda x)]u(x, t)dx + \left[\sin(\lambda x)\frac{du}{dx} - u\frac{d}{dx}[\sin(\lambda x)]\right]_0^1$$

and applying the boundary conditions and noting that

$$\frac{d^2}{dx^2}[\sin(\lambda x)] = -\lambda^2 \sin(\lambda x)$$

we have

$$-\lambda^2 \int_{x=0}^{1} \sin(\lambda x)u(x, t)dx + [u_x \sin(\lambda x) - \lambda u \cos(\lambda x)]_0^1$$

$$= -\lambda^2 U(\lambda, t) - u(1)[\lambda \cos \lambda + H \sin \lambda]$$

Defining

$$S_\lambda\{u(x, t)\} = \int_{x=0}^{1} u(x, t) \sin(\lambda x)dx = U(\lambda, t)$$

as the Fourier sine transform of $u(x, t)$ and setting

$$\lambda \cos \lambda + H \sin \lambda = 0$$

we find

$$U_t(\lambda, t) = -\lambda^2 U(\lambda, t)$$

whose solution is

$$U(\lambda, t) = Ae^{-\lambda^2 t}$$

The initial condition of the transformed function is

$$U(\lambda, 0) = \int_{x=0}^{1} \sin(\lambda x)dx = \frac{1}{\lambda}[1 - \cos(\lambda)]$$

Applying the initial condition we find

$$U(\lambda, t) = \frac{1}{\lambda}[1 - \cos(\lambda)]e^{-\lambda^2 t}$$

It now remains to find from this the value of $u(x, t)$.

Recall from the general theory of Fourier series that any odd function of x defined on $0 \le x \le 1$ can be expanded in a Fourier sine series in the form

$$u(x, t) = \sum_{n=1}^{\infty} \frac{\sin(\lambda_n x)}{\|\sin(\lambda_n)\|^2} \int_{\xi=0}^{1} u(\xi, t) \sin(\lambda_n \xi)d\xi$$

and this is simply

$$u(x, t) = \sum_{n=1}^{\infty} \frac{\sin(\lambda_n x)}{\|\sin(\lambda_n)\|^2} U(\lambda_n, t)$$

with λ_n given by the transcendental equation above. The final solution is therefore

$$u(x, t) = \sum_{n=1}^{\infty} \frac{2(1 - \cos \lambda_n)}{\lambda_n - \frac{1}{2} \sin(2\lambda_n)} \sin(\lambda_n x) e^{-\lambda_n^2 t}$$

9.2   GENERALIZATION: THE STURM–LIOUVILLE TRANSFORM: THEORY

Consider the differential operator D

$$D[f(x)] = A(x)f'' + B(x)f' + C(x)f \qquad a \le x \le b \qquad (9.1)$$

with boundary conditions of the form

$$\begin{aligned} N_\alpha[f(x)]_{x=a} &= f(a)\cos\alpha + f'(a)\sin\alpha \\ N_\beta[f(x)]_{x=b} &= f(b)\cos\beta + f'(b)\sin\beta \end{aligned} \qquad (9.2)$$

where the symbols N_α and N_β are differential operators that define the boundary conditions. For example the differential operator might be

$$D[f(x)] = f_{xx}$$

and the boundary conditions might be defined by the operators

$$N_\alpha[f(x)]_{x=a} = f(a) = 0$$

and

$$N_\beta[f(x)]_{x=b} = f(b) + Hf'(b) = 0$$

We define an integral transformation

$$T[f(x)] = \int_a^b f(x)K(x, \lambda)dx = F(\lambda) \qquad (9.3)$$

We wish to transform these differential forms into algebraic forms. First we write the differential operator in standard form. Let

$$r(x) = \exp \int_a^x \frac{B(\xi)}{A(\xi)} d\xi$$

$$p(x) = \frac{r(x)}{A(x)} \tag{9.4}$$

$$q(x) = -p(x)C(x)$$

Then

$$D[f(x)] = \frac{1}{p(x)} \left[(rf')' - qf \right] = \frac{1}{p(x)} \Re[f(x)] \tag{9.5}$$

where $\Re$ is the Sturm–Liouville operator.

Let the kernel function $K(x, \lambda)$ in Eq. (9.3) be

$$K(x, \lambda) = p(x)\Phi(x, \lambda) \tag{9.6}$$

Then

$$T[D[f(x)]] = \int_a^b \Phi(x, \lambda)\Re[f(x)]dx$$

$$= \int_a^b f(x)\Re[\Phi(x, \lambda)]dx + [(\Phi f_x - \Phi_x f)r(x)]_a^b \tag{9.7}$$

while

$$N_\alpha[f(a)] = f(a)\cos\alpha + f(a)\sin\alpha$$

$$N_\alpha'[f(a)] = \frac{d}{d\alpha}\left(f(a)\cos\alpha + f'(a)\sin\alpha\right) \tag{9.8}$$

$$= -f(a)\sin\alpha + f'(a)\cos\alpha$$

so that

$$f(a) = N_\alpha[f(a)]\cos\alpha - N_\alpha'[f(a)]\sin\alpha$$

$$f'(a) = N_\alpha'[f(a)]\cos\alpha + N_\alpha[f(a)]\sin\alpha \tag{9.9}$$

where the prime indicates differentiation with respect to α.

The lower boundary condition at $x = a$ is then

$$[\Phi(a, \lambda)f'(a) - \Phi'(a, \lambda)f(a)]\, r(a)$$
$$= \begin{bmatrix} \Phi(a, \lambda)N'_\alpha[f(a)]\cos\alpha + \Phi(a, \lambda)N_\alpha[f(a)]\sin\alpha - \Phi'(a, \lambda)N_\alpha[f(a)]\cos\alpha \\ +\Phi'(a, \lambda)N'_\alpha[f(a)]\sin\alpha \end{bmatrix} r(a)$$

$$(9.10)$$

But if $\Phi(x, \lambda)$ is chosen to satisfy the Sturm–Liouville equation and the boundary conditions then

$$N_\alpha[\Phi(x, \lambda)]_{x=a} = \Phi(a, \lambda)\cos\alpha + \Phi'(a, \lambda)\sin\alpha$$
$$N_\beta[\Phi(x, \lambda)]_{x=b} = \Phi(b, \lambda)\cos\beta + \Phi'(b, \lambda)\sin\beta$$

$$(9.11)$$

and

$$\Phi(a, \lambda) = N_\alpha[\Phi(a, \lambda)]\cos\alpha - N'_\alpha[\Phi(a, \lambda)]\sin\alpha$$
$$\Phi'(a, \lambda) = N'_\alpha[\Phi(a, \lambda)]\cos\alpha + N_\alpha[\Phi(a, \lambda)]\sin\alpha$$

$$(9.12)$$

and we have

$$[(N'_\alpha[\Phi(a, \lambda)]\cos\alpha + N_\alpha[f(a)]\sin\alpha)(N_\alpha[\Phi(a, \lambda)]\cos\alpha + N'_\alpha[\Phi(a, \lambda)]\sin\alpha)$$
$$- (N'_\alpha[\Phi(a, \lambda)]\cos\alpha + N_\alpha[\Phi(a, \lambda)]\sin\alpha)(N'_\alpha[f(a)]\cos\alpha$$
$$- N_\alpha[f(a)]\sin\alpha)]r(a)$$

$$(9.13)$$

$$= \{N'_\alpha[f(a)]N_\alpha[\Phi(a, \lambda)] - N_\alpha[f(a)]N'_\alpha[\Phi(a, \lambda)]\}r(a)$$

If the kernel function is chosen so that $N_\alpha[\Phi(a, \lambda)] = 0$, for example, the lower boundary condition is

$$-N_\alpha[f(a)]N'_\alpha[\Phi(a, \lambda)]r(a)$$

$$(9.14)$$

Similarly, at $x = b$

$$\left[\Phi(b, \lambda)f'(b) - \Phi'(b, \lambda)f(b)\right]r(b) = -N_\beta[f(b)]N'_\beta[\Phi(b, \lambda)]r(b)$$

$$(9.15)$$

Since $\Phi(x, \lambda)$ satisfies the Sturm–Liouville equation, there are n solutions forming a set of orthogonal functions with weight function $p(x)$ and

$$\Re\Phi_n(x, \lambda_n) = -\lambda_n^2 p(x)\Phi_n(x, \lambda_n)$$

$$(9.16)$$

so that

$$T\{D[f(x)]\} = -\lambda^2 \int_{x=a}^{b} p(x)f(x)\Phi_n(x, \lambda)dx + N_\alpha[f(a)]N_\alpha'[\Phi_n(a, \lambda)]r(a)$$

$$- N_\beta[f(b)]N_\beta'[\Phi_n(b, \lambda)]r(b) \qquad (9.17)$$

where

$$\lambda_n^2 \int_a^b p(x)f_n(x)\Phi_n(x, \lambda_n)dx = \lambda_n^2 F_n(\lambda_n) \qquad (9.18)$$

9.3   THE INVERSE TRANSFORM

The great thing about Sturm–Liouville transforms is that the inversion is so easy. Recall that the generalized Fourier series of a function $f(x)$ is

$$f(x) = \sum_{n=1}^{\infty} \frac{\Phi_n(x, \lambda_n)}{\|\Phi_n\|} \int_a^b f_n(\xi)p(\xi)\frac{\Phi_n(\xi, \lambda_n)}{\|\Phi_n\|}d\xi = \sum_{n=1}^{\infty} \frac{\Phi_n(x)}{\|\Phi_n\|^2}F(\lambda_n) \qquad (9.19)$$

where the functions $\Phi_n(x, \lambda_n)$ form an orthogonal set with respect to the weight function $p(x)$.

Example 9.2 (The cosine transform).   Consider the diffusion equation

$$y_t = y_{xx} \qquad 0 \le x \le 1 \qquad t > 0$$

$$y_x(0, t) = y(1, t) = 0$$

$$y(x, 0) = f(x)$$

To find the proper kernel function $K(x, \lambda)$ we note that according to Eq. (9.16) $\Phi_n(x, \lambda_n)$ must satisfy the Sturm–Liouville equation

$$\Re[\Phi_n(x, \lambda)] = -p(x)\Phi_n(x, \lambda)$$

where for the current problem

$$\Re[\Phi_n(x, \lambda)] = \frac{d^2}{dx^2}[\Phi_n(x, \lambda)] \quad \text{and} \quad p(x) = 1$$

along with the boundary conditions (9.11)

$$N_\alpha[\Phi(x, \lambda)]_{x=a} = \Phi_x(0, \lambda) = 0$$

$$N_\beta[\Phi(x, \lambda)]_{x=b} = \Phi(1, \lambda) = 0$$

Solution of this differential equation and applying the boundary conditions yields an infinite number of functions (as any Sturm–Liouville problem)

$$\Phi(x, \lambda_n) = A\cos(\lambda_n x)$$

with

$$\cos(\lambda_n) = 0 \qquad \lambda_n = \frac{(2n-1)}{2}\pi$$

Thus, the appropriate kernel function is $K(x, \lambda_n) = \cos(\lambda_n x)$ with $\lambda_n = (2n-1)\frac{\pi}{2}$. Using this kernel function in the original partial differential equation, we find

$$\frac{dY}{dt} = -\lambda_n^2 Y$$

where $C_\lambda\{y(x, t)\} = Y(t, \lambda_n)$ is the cosine transform of $y(t, x)$. The solution gives

$$Y(t, \lambda_n) = Be^{-\lambda^2 t}$$

and applying the cosine transform of the initial condition

$$B = \int_{x=0}^{1} f(x)\cos(\lambda_n x)dx$$

According to Eq. (9.19) the solution is as follows:

$$y(x, t) = \sum_{n=0}^{\infty} \frac{\cos(\lambda_n x)}{\|\cos(\lambda_n x)\|^2} \int_{x-0}^{1} f(x)\cos(\lambda_n x)dx e^{-\lambda_n^2 t}$$

Example 9.3 (The Hankel transform).   Next consider the diffusion equation in cylindrical coordinates.

$$u_t = \frac{1}{r}\frac{d}{dr}\left(r\frac{du}{dr}\right)$$

Boundary and initial conditions are prescribed as

$$u_r(t, 0) = 0$$
$$u(t, 1) = 0$$
$$u(0, r) = f(r)$$

First we find the proper kernel function

$$\Re[\Phi(r, \lambda_n)] = \frac{d}{dr}\left(r\frac{d\Phi_n}{dr}\right) = -\lambda_n^2 r\,\Phi$$

with boundary conditions

$$\Phi_r(\lambda_n, 0) = 0$$

$$\Phi(\lambda_n, 1) = 0$$

The solution is the Bessel function $J_0(\lambda_n r)$ with λ_n given by $J_0(\lambda_n) = 0$. Thus the transform of $u(t, r)$ is as follows:

$$H_\lambda \{u(t, r)\} = U(t, \lambda_n) = \int_{r=0}^{1} r J_0(\lambda_n r) u(t, r) dr$$

This is called a *Hankel transform*. The appropriate differential equation for $U(t, \lambda_n)$ is

$$\frac{dU_n}{dt} = -\lambda_n^2 U_n$$

so that

$$U_n(t, \lambda_n) = Be^{-\lambda_n^2 t}$$

Applying the initial condition, we find

$$B = \int_{r=0}^{1} r f(r) J_0(\lambda_n r) dr$$

and from Eq. (9.19)

$$u(t, r) = \sum_{n=0}^{\infty} \frac{\int_{r=0}^{1} r f(r) J_0(\lambda_n r) dr}{\left\| J_0(\lambda_n r) \right\|^2} J_0(\lambda_n r) e^{-\lambda_n^2 t}$$

Example 9.4 (The sine transform with a source).   Next consider a one-dimensional transient diffusion with a source term $q(x)$:

$$u_t = u_{xx} + q(x)$$

$$y(0, x) = y(t, 0) = t(t, \pi) = 0$$

First we determine that the sine transform is appropriate. The operator $\Re$ is such that

$$\Re\Phi = \Phi_{xx} = \lambda\Phi$$

and according to the boundary conditions we must choose $\Phi = \sin(nx)$ and $\lambda = -n^2$. The sine transform of $q(x)$ is $Q(\lambda)$.

$$U_t = -n^2 U + Q(\lambda)$$
$$U = U(\lambda, t)$$

The homogeneous and particular solutions give

$$U_n = Ce^{-n^2 t} + \frac{Q_n}{n^2}$$

when $t = 0$, $U = 0$ so that

$$C = -\frac{Q_n}{n^2}$$

where Q_n is given by

$$Q_n = \int\limits_{x=0}^{\pi} q(x)\sin(nx)dx$$

Since $U_n = \frac{Q_n}{n^2}[1 - e^{-n^2 t}]$ the solution is

$$u(x, t) = \sum_{n=1}^{\infty} \frac{Q_n}{n^2}[1 - e^{-n^2 t}]\frac{\sin(nx)}{\left\|\sin(nx)\right\|^2}$$

Note that Q_n is just the nth term of the Fourier sine series of $q(x)$. For example, if $q(x) = x$,

$$Q_n = \frac{\pi}{n}(-1)^{n+1}$$

Example 9.5 (A mixed transform). Consider steady temperatures in a half cylinder of infinite length with internal heat generation, $q(r)$ that is a function of the radial position. The appropriate differential equation is

$$u_{rr} + \frac{1}{r}u_r + \frac{1}{r^2}u_{\theta\theta} + u_{zz} + q(r) = 0 \quad 0 \le r \le 1 \quad 0 \le z \le \infty \quad 0 \le \theta \le \pi$$

with boundary conditions

$$u(1, \theta, z) = 1$$
$$u(r, 0, z) = u(r, \pi, z) = u(r, \theta, 0) = 0$$

Let the sine transform of u be denoted by $S_n\{u(r, \theta, z)\} = U_n(r, n, z)$ with respect to θ on the interval $(0, \pi)$. Then

$$\frac{\partial^2 U_n}{\partial r^2} + \frac{1}{r}\frac{\partial U_n}{\partial r} - \frac{n^2}{r^2}U_n + \frac{\partial^2 U_n}{\partial z^2} + q(r)S_n(1) = 0$$

where $S_n(1)$ is the sine transform of 1, and the boundary conditions for $u(r, \theta, z)$ on θ have been used.

Note that the operator on Φ in the r coordinate direction is

$$\Re\left[\Phi(r, \mu_j)\right] = \frac{1}{r}\frac{d}{dr}\left(r\frac{d\Phi}{dr}\right) - \frac{n^2}{r^2}\Phi = -\mu_j^2\Phi$$

With the boundary condition at $r = 1$ chosen as $\Phi(1, \mu_j) = 0$ this gives the kernel function as $\Phi = rJ_n(r, \mu_j)$ with eigenvalues determined by $J_n(1, \mu_j) = 0$

We now apply the finite Hankel transform to the above partial differential equation and denote the Hankel transform of U_n by U_{jn}.

After applying the boundary condition on r we find, after noting that

$$N_\beta[U_n(z, 1)] = S_n(1)$$

$$N_\beta'[\Phi(1, z)] = -\mu_j J_{n+1}(\mu_j)$$

$-\mu_j^2 U_{jn} + \mu_j J_{n+1}(\mu_j)S_n(1) + \frac{d^2 U_{jn}}{dz^2} + Q_j(\mu_j)S_n(1) = 0$. Here $Q_j(\mu_j)$ is the Hankel transform of $q(r)$.

Solving the resulting ordinary differential equation and applying the boundary condition at $z = 0$,

$$U_{jn}(\mu_j, n, z) = S_n(1)\frac{Q_j(\mu_j) + \mu_j J_{n+1}(\mu_j)}{\mu_j^2}[1 - \exp(-\mu_j z)]$$

We now invert the transform for the sine and Hankel transforms according to Eq. (9.19) and find that

$$u(r, \theta, z) = \frac{4}{\pi}\sum_{n=1}^{\infty}\sum_{j=1}^{\infty}\frac{U_{jn}(\mu_j, n, z)}{[J_{n+1}(\mu_j)]^2}J_n(\mu_j r)\sin(n\theta)$$

Note that

$$S_n(1) = [1 - (-1)^n]/n$$

Problems

Use an appropriate Sturm–Liouville transform to solve each of the following problems:

1. Chapter 3, Problem 1.
2. Chapter 2, Problem 2.
3. Chapter 3, Problem 3.

$$\frac{\partial u}{\partial t} = \frac{1}{r}\frac{\partial}{\partial r}\left(r\frac{\partial u}{\partial r}\right) + G(\text{constant } t)$$

4. $u(r, 0) = 0$

    $u(1, t) = 0$

    u bounded

5. Solve the following using an appropriate Sturm–Liouville transform:

$$\frac{\partial^2 u}{\partial x^2} = \frac{\partial u}{\partial t}$$

$$u(t, 0) = 0$$

$$u(t, 1) = 0$$

$$u(0, x) = \sin(\pi x)$$

6. Find the solution for general $\rho(t)$:

$$\frac{\partial u}{\partial t} = \frac{\partial^2 u}{\partial x^2}$$

$$u(t, 0) = 0$$

$$u(t, 1) = \rho(t)$$

$$u(0.x) = 0$$

FURTHER READING

V. S. Arpaci, *Conduction Heat Transfer*, Reading, MA: Addison-Wesley, 1966.

R. V. Churchill, *Operational Mathematics*, 3rd ed. New York: McGraw-Hill, 1972.

I. H. Sneddon, *The Use of Integral Transforms*, New York: McGraw-Hill, 1972.

C H A P T E R   10

Introduction to Perturbation Methods

Perturbation theory is an approximate method of solving equations which contain a parameter that is small in some sense. The method should result in an approximate solution that may be termed "precise" in the sense that the error (the difference between the approximate and exact solutions) is understood and controllable and can be made smaller by some rational technique. Perturbation methods are particularly useful in obtaining solutions to equations that are nonlinear or have variable coefficients. In addition it is important to note that if the method yields a simple, accurate approximate solution of any problem it may be more useful than an exact solution that is more complicated.

10.1   EXAMPLES FROM ALGEBRA

We begin with examples from algebra in order to introduce the ideas of regular perturbations and singular perturbations. We start with a problem of extracting the roots of a quadratic equation that contains a small parameter $\varepsilon \ll 1$.

10.1.1   REGULAR PERTURBATION

Example 10.1   Consider, for example the equation

$$x^2 + \varepsilon x - 1 = 0 \tag{10.1}$$

The exact solution for the roots is, of course, simply obtained from the quadratic formula.

$$x = -\frac{\varepsilon}{2} \pm \sqrt{1 + \frac{\varepsilon^2}{4}} \tag{10.2}$$

which for $\varepsilon = 0.1$ yields exact solutions

$$x = 0.962422837$$
$$\text{and}$$
$$x = -1.062422837$$

Eq. (10.2) can be expanded for small values of ε in the rapidly convergent series

$$x = 1 - \frac{\varepsilon}{2} + \frac{\varepsilon^2}{8} - \frac{\varepsilon^4}{128} + \ldots \ldots \tag{10.3}$$

or

$$x = -1 - \frac{\varepsilon}{2} - \frac{\varepsilon^2}{8} + \frac{\varepsilon^4}{128} - \ldots \ldots \tag{10.4}$$

To apply perturbation theory we first note that if $\varepsilon = 0$ the two roots of the equation, which we will call the zeroth order solutions, are $x_0 = \pm 1$. We assume a solution of the form

$$x = x_0 + a_1\varepsilon + a_2\varepsilon^2 + a_3\varepsilon^3 + a_4\varepsilon^4 + \ldots \tag{10.5}$$

Substituting (10.5) into (10.1)

$$1 + (2a_1 + 1)\varepsilon + (a_1^2 + 2a_2 + a_1)\varepsilon^2 + (2a_1a_2 + 2a_3 + a_2)\varepsilon^3 + \ldots \ldots - 1 = 0 \tag{10.6}$$

where we have substituted $x_0 = 1$. Each of the coefficients of ε^n must be zero. Solving for a_n we find

$$\begin{aligned} a_1 &= -\tfrac{1}{2} \\ a_2 &= \tfrac{1}{8} \\ a_3 &= 0 \end{aligned} \tag{10.7}$$

So that the approximate solution for the root near $x = 1$ is

$$x = 1 - \frac{\varepsilon}{2} + \frac{\varepsilon^2}{8} + O\left(\varepsilon^4\right) \tag{10.8}$$

The symbol $O(\varepsilon^4)$ means that the next term in the series is of order ε^4.

Performing the same operation with $x_0 = -1$

$$1 - (1 + 2a_1)\varepsilon + (a_1^2 - 2a_2 + a_1)\varepsilon^2 + (2a_1a_2 - 2a_3 + a_2)\varepsilon^3 + \ldots - 1 = 0 \tag{10.9}$$

Again setting the coefficients of ε^n equal to zero

$$\begin{aligned} a_1 &= -\tfrac{1}{2} \\ a_2 &= -\tfrac{1}{8} \\ a_3 &= 0 \end{aligned} \tag{10.10}$$

so that the root near $x_0 = -1$ is

$$x = -1 - \frac{\varepsilon}{2} - \frac{\varepsilon^2}{8} + O\left(\varepsilon^4\right) \tag{10.11}$$

The first three terms in (10.8) give $x = 0.951249219$, accurate to within 1.16% of the exact value while (10.11) gives the second root as $x = -1.051249219$, which is accurate to within 1.05%.

Example 10.2   Next suppose the small parameter occurs multiplied by the squared term.

$$\varepsilon x^2 + x - 1 = 0 \tag{10.12}$$

Using the quadratic formula gives the exact solution.

$$x = -\frac{1}{2\varepsilon} \pm \sqrt{\frac{1}{4\varepsilon^2} + \frac{1}{\varepsilon}} \tag{10.13}$$

If $\varepsilon = 0.1$ (10.13) gives two solutions:

$$x = 0.916079783$$
$$\text{and}$$
$$x = -10.91607983$$

We attempt to follow the same procedure that we used in Example 10.1, which we call a *regular perturbation*, to obtain an approximate solution. If $\varepsilon = 0$ identically, $x_0 = 1$. Using (10.5) with $x_0 = 1$ and substituting into (10.12) we find

$$(1 + a_1)\varepsilon + (2a_1 + a_2)\varepsilon^2 + (2a_2 + a_1^2 + a_3)\varepsilon^3 + ... = 0 \tag{10.14}$$

Setting the coefficients of $\varepsilon^n = 0$, solving for a_n and substituting into (10.5)

$$x = 1 - \varepsilon + 2\varepsilon^2 - 5\varepsilon^3 + ... \tag{10.15}$$

gives $x = 0.915$, close to the exact value. However, Eq. (10.12) clearly has two roots, and the method cannot give an approximation for the second root.

The essential problem is that the second root is not small. In fact (10.13) shows that as $\varepsilon \to 0$, $|x| \to \frac{1}{2\varepsilon}$ so that the term εx^2 is never negligible.

10.1.2 SINGULAR PERTURBATION

Arranging (10.12) in normal form

$$x^2 + \frac{x - 1}{\varepsilon} = 0 \tag{10.12a}$$

and the equation is said to be singular as $\varepsilon \to 0$. If we set $x\varepsilon = u$ we find an equation for u as

$$u^2 + u - \varepsilon = 0 \tag{10.16}$$

With ε identically zero, $u = 0$ or -1. Assuming that u may be approximated by a series like (10.5) we find that

$$(-a_1 - 1)\varepsilon + (a_1^2 - a_2)\varepsilon^2 + (2a_1a_2 - a_3)\varepsilon^3 + ... = 0 \tag{10.17}$$

$$a_1 = -1$$
$$a_2 = 1 \tag{10.18}$$
$$a_3 = -2$$

so that

$$x = -\frac{1}{\varepsilon} - 1 + \varepsilon - 2\varepsilon^2 + \dots\dots \tag{10.19}$$

The three term approximation of the negative root is therefore $x = -10.92$, within .03% of the exact solution.

Example 10.3   As a third algebraic example consider

$$x^2 - 2\varepsilon x - \varepsilon = 0 \tag{10.20}$$

This at first seems like a harmless problem that appears at first glance to be amenable to a regular perturbation expansion since the x^2 term is not lost when $\varepsilon \to 0$. We proceed optimistically by taking

$$x = x_0 + a_1\varepsilon + a_2\varepsilon^2 + a_3\varepsilon^3 + \dots \tag{10.21}$$

Substituting into (10.20) we find

$$x_0^2 + (2x_0 a_1 - 2x_0 - 1)\varepsilon + (a_1^2 + 2x_0 a_2 - 2a_1)\varepsilon^2 + \dots = 0 \tag{10.22}$$

from which we find

$$\begin{aligned} x_0 &= 0 \\ 2x_0 a_1 - 2x_0 - 1 &= 0 \\ a_1^2 + 2x_0 a_2 - 2a_1 &= 0 \end{aligned} \tag{10.23}$$

From the second of these we conclude that either $x_0 = -1$ or that there is something wrong. That is, (10.21) is not an appropriate expansion in this case.

Note that (10.20) tells us that as $\varepsilon \to 0$, $x \to 0$. Moreover, in writing (10.21) we have essentially assumed that $\varepsilon \to 0$ in such a manner that $\frac{x}{\varepsilon} \to$ constant. Let us suppose instead that as $\varepsilon \to 0$

$$\frac{x(\varepsilon)}{\varepsilon^p} \to \text{constant} \tag{10.24}$$

We then define a new variable

$$x = \varepsilon^p v(\varepsilon) \tag{10.25}$$

such that $v(0) \neq 0$. Substitution into (10.20) yields

$$\varepsilon^{2p} v^2 - 2\varepsilon^{p+1} v - \varepsilon = Q \tag{10.26}$$

where Q must be *identically* zero. Note that $\frac{Q}{\varepsilon}$ must also be zero no matter how small ε becomes, as long as it is not identically zero.

Now, if $p > \frac{1}{2}$, $2p - 1 > 0$ and in the limit as $\varepsilon - 0$, $\varepsilon^{2p-1} v^2(\varepsilon) - 2\varepsilon^p v(\varepsilon) - 1 \to -1$, which cannot be true given that $Q = 0$ identically.

Next suppose $p < 1/2$. Again, $\frac{Q}{\varepsilon^{2p}}$ is identically zero for all ε including in the limit as $\varepsilon \to 0$. In the limit as $\varepsilon \to 0$, $v(\varepsilon)^2 - \varepsilon^{1-p}v(\varepsilon) - \varepsilon^{1-2p} \to v(0)^2 \neq 0$. Thus, $p = 1/2$ is the only possibility left, so we attempt a solution with this value. Hence,

$$x = \varepsilon^{1/2}v(\varepsilon) \tag{10.27}$$

Substitution into (10.20) gives

$$v^2 - 2\sqrt{\varepsilon}v - 1 = 0 \tag{10.28}$$

and this can now be solved by a regular perturbation assuming $\beta = \sqrt{\varepsilon} << 1$. Hence,

$$v = v_0 + a_1\beta + a_2\beta^2 + a_3\beta^3 + \ldots \tag{10.29}$$

Inserting this into (10.28) with $\beta = \sqrt{\varepsilon}$

$$v_0 - 1 + (2v_0 a_1 - 2v_0)\beta + (a_1^2 + 2v_0 a_2 - 2a_1)\beta^2 + \ldots = 0 \tag{10.30}$$

Thus,

$$\begin{aligned} v_0 &= \pm 1 \\ a_1 &= 1 \\ a_2 &= +\tfrac{1}{2} \quad \text{or} \quad -\tfrac{1}{2} \end{aligned} \tag{10.31}$$

Thus, the two solutions are

$$v = 1 + \sqrt{\varepsilon} + \frac{\varepsilon}{2} + \ldots \tag{10.32}$$

and

$$v = -1 + \sqrt{\varepsilon} - \frac{\varepsilon}{2} + \ldots \tag{10.33}$$

The approximate solutions are

$$x = \sqrt{\varepsilon} + \varepsilon + \frac{1}{2}\varepsilon\sqrt{\varepsilon} + \ldots \tag{10.34}$$

and

$$x = -\sqrt{\varepsilon} + \varepsilon - \frac{1}{2}\varepsilon\sqrt{\varepsilon} + \ldots \tag{10.35}$$

If $\beta = 0.1$, approximate roots are 0.1105 and -0.0805 whereas the exact solution is, .12499 and $-.0805$. Note that when $\beta = 0.1$, $\varepsilon = 0.36623$, so that ε is not particularly small.

Example 10.4   Finally, consider the third order algebraic equation,

$$\varepsilon x^3 + x - 2 = 0 \tag{10.36}$$

While the exact solution of the quadratic equation is easy to find, the solution to the cubic is not so easy. Let us see what singular perturbation theory can provide as an approximation.

First we try a regular perturbation expansion of the form

$$x = x_0 + x_1 \varepsilon + \varepsilon^2 x_2 + ...$$

(10.37)

Substituting into Eq. (10.36) gives the following set of equations:

$$\begin{aligned} \varepsilon^0 &: x_0 = 2 \\ \varepsilon^1 &: x_1 = -x_0^3 \\ \varepsilon^2 &: x_2 = -3x_0^2 x_1 \end{aligned}$$

(10.38)

Thus, the regular perturbation yields only one root, which is approximated by

$$x = 2 - 8\varepsilon + 96\varepsilon^2 + O\left(\varepsilon^3\right)$$

(10.39)

The problem is that, once again, as $\varepsilon \to 0$ the first term is lost unless x becomes very large. But if x is very large, what must we do about the second term? Let us again explore what happens if we let

$$x = \varepsilon^p v(\varepsilon)$$

(10.40)

Equation (10.36) becomes

$$\varepsilon^{1+3p} v^3(\varepsilon) + \varepsilon^p v(\varepsilon) - 2 = 0$$

(10.41)

As $\varepsilon \to 0$ the first two terms retain the same orders of magnitude only if $p = -1/2$. In this case,

$$v^3 + v - 2\varepsilon^{1/2} = 0$$

(10.42)

We try an expansion of the form

$$v = \left(a_0 + \varepsilon^{1/2} a_1 + \varepsilon a_2 + ...\right)$$

(10.43)

Substituting into ....,

$$\left(a_0^3 + a_0\right) + \varepsilon^{1/2}\left(3a_1 a_0^2 + a_1 - 2\right) + \varepsilon\left(3a_0^2 a_2 + 3a_1^2 a_0 + a_2\right) + O\left(\varepsilon^{3/2}\right) = 0$$

(10.44)

Equating terms of the same order,

$$\begin{aligned} a_0 &= 0 \text{ or } \pm i \\ a_1 &= -1 \\ a_2 &= \pm \tfrac{3}{2} i \end{aligned}$$

(10.45)

The yields two complex roots:

$$\begin{aligned} & i\left(\varepsilon^{-1/2} + \tfrac{3}{2}\varepsilon^{1/2}\right) - 1 \\ & -i\left(\varepsilon^{-1/2} + \tfrac{3}{2}\varepsilon^{1/2}\right) - 1 \end{aligned}$$

(10.46)

Problems

Using appropriate perturbation analysis and for small ε

1. Find the approximate roots of

$$x^2 - \varepsilon x - 1 = 0$$

2. Find the approximate roots of

$$\varepsilon x^2 + 2x + 1 = 0$$

3. Find the approximate roots of

$$\varepsilon x^3 - x^2 + 1 = 0$$

and evaluate for $\varepsilon = 0.1$ and $\varepsilon = .01$

Answer:

$$\varepsilon = .1, -0.995, 1.057, 9.898$$
$$\varepsilon = .01, -0.995, 1.005, 99.990$$

10.2   EXAMPLES FROM ORDINARY DIFFERENTIAL EQUATIONS

We now introduce some ideas about approximate solutions of ordinary differential equations using perturbation methods. We first consider two examples of problems that are solvable by regular perturbation methods. Then we illustrate the failure of the regular perturbation method to yield an approximate solution in two simple equations before moving on in Chapter 11 to more fully discuss singular methods.

Example 10.5   Without going into the details of the physical derivation, the following equation and initial condition describe the cooling by convection of a small object that has a heat capacity that varies slightly with temperature.

$$(1 + \varepsilon x)\frac{dx}{dt} + x = 0$$
$$x(0) = 1$$
(10.47)

(The physical heat transfer derivation can be found in the book by Aziz and Na at the end of this chapter.)

One might think of x as a dimensionless temperature departure from that of the environment. The initial dimensionless temperature is 1 and ε is a small dimensionless parameter. As is usual in the regular perturbation approach we begin by taking

$$x = x_0 + \varepsilon x_1 + \varepsilon^2 x_2 + O\left(\varepsilon^3\right)$$
(10.48)

Inserting this into (10.47) results in

$$\frac{dx_0}{dt} + x_0 + \varepsilon \left(\frac{dx_1}{dt} + x_1 + x_0 \frac{dx_0}{dt} \right) + \varepsilon^2 \left(\frac{dx_2}{dt} + x_2 + x_0 \frac{dx_1}{dt} + x_1 \frac{dx_0}{dt} \right) + O\left(\varepsilon^3 \right) = 0 \tag{10.49}$$

Thus,

$$\frac{dx_0}{dt} + x_0 = 0, \, x_0(0) = 1$$

$$\frac{dx_1}{dt} + x_1 + x_0 \frac{dx_0}{dt} = 0, \, x_1(0) = 0 \tag{10.50}$$

$$\frac{dx_2}{dt} + x_2 + x_0 \frac{dx_1}{dt} + x_1 \frac{dx_0}{dt} = 0. \, \, x_2(0) = 0$$

The solution of the first of these is

$$x_0 = e^{-t} \tag{10.51}$$

The second equation (associated with the ε term) is

$$x_1 = e^{-t} - e^{-2t} \tag{10.52}$$

In the case of the third equation we note that

$$x_2 = e^{-t} - 2e^{-2t} + e^{-3t} \tag{10.53}$$

The approximate solution is then

$$x = e^{-t} + \varepsilon \left(e^{-t} - e^{-2t} \right) + \varepsilon^2 \left(e^{-t} - 2e^{-2t} + e^{-3t} \right) + O\left(\varepsilon^3 \right) \tag{10.54}$$

Note that the equation can be easily solved by separation of variables and the exact solution is

$$\ln x + \varepsilon (x - 1) = -t \tag{10.55}$$

(We comment here that in (10.55), x is a rather complicated implicit function of t. Equations like (10.54) are often easier to work with and, as long as they are sufficiently accurate, preferred. The author strongly feels that a good approximate solution that is simple is often more useful than a complicated exact solution.)

Example 10.6   Next consider the case of a spring mass damper system with a small spring constant. The mass is initially displaced by a non-dimensional amount $x(0) = 1$ and with a non-dimensional velocity $\frac{dx}{dt}(0) = 0$. The differential equation and initial conditions are as follows:

$$\frac{d^2 x}{dt_2} + \frac{dx}{dt} + \varepsilon x = 0$$

$$x(0) = 1 \tag{10.56}$$

$$\frac{dx}{dt}(0) = 0$$

We attempt a regular perturbation solution in the form

$$x = x_0 + \varepsilon x_1 + \varepsilon^2 x_2 + O\left(\varepsilon^3\right) \tag{10.57}$$

Substituting in the usual way we find

$$\frac{d^2 x_0}{dt^2} + \frac{dx_0}{dt} + \varepsilon\left(\frac{d^2 x_1}{dt^2} + \frac{dx_1}{dt} + x_0\right) + O\left(\varepsilon^2\right) = 0$$

$$x_0(0) = 1, \quad \frac{dx_0}{dt}(0) = 0 \tag{10.58}$$

$$x_1(0) = \frac{dx_1}{dt}(0) = 0$$

(It is clear to the reader at this point that it is easy to solve the original differential equation, since it is linear and rather easy to solve by Laplace transforms or other means. We continue merely to illustrate the perturbation method.)

Solving,

$$x_0 = 1$$
$$x_1 = 2 - t - e^{-t} \tag{10.59}$$

We can see a problem arising here that often arises when using approximate methods. As long as t is sufficiently small, the solution,

$$x = 1 - \varepsilon\left(t + e^{-t} - 2\right) + O\left(\varepsilon^2\right) \tag{10.60}$$

is accurate. However, when t is larger than, perhaps, ε, the result is not useful. In fact, when t approaches infinity, x approaches negative infinity, which clearly cannot be true. Keeping the ε^2 term helps, but not much.

Example 10.7 Next consider the case of the mass, spring damper system in which the mass is small. Initially the mass is at rest, $x(0) = \frac{dx}{dt} = 0$, and a sudden force, a dimensionless impulsive force of $F(0) = \delta(0)$, is applied. We represent this force as $1/\varepsilon$. The appropriate differential equation and initial conditions are

$$\varepsilon\frac{d^2 x}{dt^2} + \frac{dx}{dt} + x = 0$$

$$x(0) = 0 \tag{10.61}$$

$$\frac{dx}{dt}(0) = 1/\varepsilon$$

Physically, when ε is very small, this corresponds to an impulsive force acting on the mass, giving it a large initial speed. From experience with the algebraic equations, we might expect that, with the small term multiplying the second derivative term, a regular perturbation approach might not work. Such will herewith be demonstrated.

Begin by assuming a regular perturbation approach with

$$x = x_0 + \varepsilon x_1 + O\left(\varepsilon^2\right) \tag{10.62}$$

$$\varepsilon \frac{d^2 x_0}{dt^2} + \varepsilon^2 \frac{d^2 x_1}{dt^2} + \frac{dx_0}{dt} + \varepsilon \frac{dx_1}{dt} + x_0 + \varepsilon x_1 = 1 \tag{10.63}$$

The first order equation, the equation when $\varepsilon = 0$, is

$$\frac{dx_0}{dt} + x_0 = 0 \tag{10.64}$$

The second derivative term has been eliminated, so that the solution cannot satisfy the initial conditions, which require both the terms on the left hand side of the equation to be zero at $t = 0$. Clearly, the regular perturbation approach will not work.

We now turn to an exposition of the classical use of singular perturbation methods in the useful approximate solutions of boundary value problems.

Problems

Use regular perturbation theory to find an approximate solution for small ε

10.4. $y'' + \varepsilon y' + 1 = 0, \ y(0) = 0, \ y'(0) = 1$

10.5. $y' + y + \varepsilon y^2 = x, \ y(1) = 1$

REFERENCES

[1] Aziz, A. and T. Y. Na, *Perturbation Methods in Heat Transfer*, Hemisphere Pub. Co., New York. 1984. Cited on page(s)

[2] Simmonds, J. G. and J. E. Mann, *A First Look at Perturbation Theory*, Robert E. Krieger Pub. Co., Melebar, Florida, 1986. Cited on page(s)

CHAPTER 11

Singular Perturbation Theory of Differential Equations

A singular perturbation approach is usually required when a small coefficient is multiplied by the highest derivative term, so that in some region of the domain, the dependent variable changes so rapidly that this term, although multiplied by a small parameter, cannot be considered to be negligible. This often (but not always) occurs at a boundary of the domain. The approach is usually the following:

The independent variable, say t, is replaced by a variable that depends on the small parameter, say ε, by

$$\eta = \vartheta(\varepsilon) x \tag{11.1}$$

and $\varphi(\varepsilon)$ is chosen to remove the small parameter from the highest derivative term. This should allow a regular perturbation expansion to be performed within the region in which the independent variable changes rapidly, which is called the *inner region*. Next, a regular perturbation is performed with the result generally being appropriate for the region outside that where the dependent variable changes rapidly. This is the *outer region*. An intermediate region is now identified in which the inner and outer solutions form a *common solution* in some sense (best demonstrated by the examples below). This is referred to as the *matching problem*. A *uniformly valid solution* can now be obtained as the sum of the inner and outer solutions minus the common solution.

We begin our illustrative examples by reconsidering the problem introduced in Example 10.7 in the last chapter, which is often used as an introduction to some of the fundamental problems in the concept of singular perturbation theory.

Example 11.1 We begin with Eq. (10.61). The regular perturbation method has failed because the first approximation yielded a first order differential equation whose boundary conditions could not both be satisfied. If we change the independent variable to

$$t = \varepsilon \tau \tag{11.2}$$

Equation (10.61) becomes

$$\frac{d^2x}{d\tau^2} + \frac{dx}{d\tau} + \varepsilon x = 0$$

$$x(0) = 0 \tag{11.3}$$

$$\frac{dx}{d\tau} = 1$$

Now let

$$x = x_0 + \varepsilon x_1 + O\left(\varepsilon^2\right) \tag{11.4}$$

Equating terms multiplied by powers of ε yields a first approximation

$$\frac{d^2 x_0}{d\tau^2} + \frac{dx_0}{d\tau} = 0$$

$$x_0(0) = 0 \tag{11.5}$$

$$\frac{dx_0}{d\tau}(0) = 1$$

The solution is

$$x_0 = 1 - e^{-\tau} \tag{11.6}$$

The differential equation for x_1 is

$$\frac{d^2 x_1}{d\tau^2} + \frac{dx_1}{d\tau} + x_0 = 0$$

$$\frac{d^2 x_1}{d\tau^2} + \frac{dx_1}{d\tau} = e^{-\tau} - 1 \tag{11.7}$$

$$x_1(0) = \frac{dx_1}{d\tau} = 0$$

Solution gives

$$x_1 = 2\left(1 - e^{-\tau}\right) - \tau\left(1 + e^{-\tau}\right) \tag{11.8}$$

The solution is good for $t = \varepsilon\tau$ small. It is called the *inner solution*.

When t is large, such that τ is of order $1/\varepsilon$ or larger, the second derivative becomes increasingly negligible, so that a regular perturbation expansion is appropriate. Hence, referring back to (10.63) and (10.64), and solving for the first two approximations,

$$x_0 = A_0\, e^{-t}$$

$$x_1 = A_1 e^{-t} - A_0 e^{-t} \tag{11.9}$$

This is referred to as the *outer solution*, the solution that is appropriate when t is large.

It is possible to obtain a *uniformly valid solution* that satisfies both the inner solution and the outer solution. This is called the *matching problem*.

We proceed in the following manner.

It seems reasonable that the inner and outer solutions should agree in some since in some overlap region that is intermediate between the inner and outer regions. If t is of order $t = O(\varepsilon)$ then t is within the inner solution region, while if t is of order $t = O(1)$ the t is within the outer solution region. We might expect, then, that an overlap region will be characterized by values of t for which $t = O(\sqrt{\varepsilon})$. Since $\sqrt{\varepsilon}$ approaches zero less rapidly than does ε, we introduce an intermediate time scale

$$\eta = \frac{t}{\sqrt{\varepsilon}} \tag{11.10}$$

To obtain a solution that is common between the inner and outer solutions we require that

$$\frac{\lim}{\varepsilon \to 0} x_{\text{inner}} \left(\sqrt{\varepsilon} \eta \right) = \frac{\lim}{\varepsilon \to 0} x_{\text{outer}} \left(\sqrt{\varepsilon} \eta \right) \tag{11.11}$$

This results in

$$\frac{\lim}{\varepsilon \to 0} \left(1 - e^{\eta / \sqrt{\varepsilon}} \right) = \frac{\lim}{\varepsilon \to 0} A_0 e^{\eta \sqrt{\varepsilon}}, \quad \text{so that} \quad A_0 = 1 \tag{11.12}$$

The uniform solution is obtained by summing the inner and outer solutions and subtracting the overlap.

$$x = 1 - e^{-t/\varepsilon} + e^{-t} - 1 = e^{-t} - e^{-t/\varepsilon} \tag{11.13}$$

Obtaining the second approximation is left as an exercise.

Example 11.2   Next consider the differential equation

$$\varepsilon\, u'' - (2 - x^2)\, u = -1, \quad 0 \le x \le 1$$

$$\varepsilon \langle \langle 1 \tag{11.14}$$

$$u(0) = 0, \quad u(1) = 1$$

It should be clear to the reader that a regular perturbation will not produce a useful solution because the small term ε is multiplied by the second derivative term, which would be eliminated if the regular perturbation were used, so the boundary conditions could not be satisfied. We search for a useful singular perturbation approach.

Let

$$x = \varphi(\varepsilon)\, \varsigma$$

$$\frac{\varepsilon}{\vartheta^2} \frac{d^2 u}{d\varsigma^2} - \left(2 - \vartheta^2 \varsigma^2 \right) u = -1 \tag{11.15}$$

If we choose $\vartheta = \varepsilon^{1/2}$ the differential equation becomes

$$\frac{d^2 u}{d\varsigma^2} - (2 - \varepsilon \varsigma^2)\, u = -1 \tag{11.16}$$

with boundary conditions

$$u(0) = 0, \quad u\left(\varsigma = \varepsilon^{-1/2} \right) = 1 \tag{11.17}$$

For the inner solution

$$u_{\text{inner}} = f_0 + \varepsilon f_1 + O\left(\varepsilon^2 \right) \tag{11.18}$$

$$f_0'' + \varepsilon f_1'' + \ldots - 2 f_0 + 2\varepsilon \varsigma^2 f_0 - 2\varepsilon f_1 + \ldots = -1 \tag{11.19}$$

Thus,

$$f_0'' - 2 f_0 = -1 \tag{11.20}$$

This is a two-point boundary value problem, so we must be careful. The solution for f_0 needs only to satisfy the boundary condition at $\varsigma = 0$.

The general solution for f_0 is

$$f_0 = \frac{1}{2} + Ae^{\sqrt{2}\varsigma} + Be^{-\sqrt{2}\varsigma} \tag{11.21}$$

To satisfy the boundary condition at $x = 0$ we must have $A + B = -1/2$.

For the solution when the second derivative term in the original equation is small (i.e., when x is large enough that $\sqrt{\varepsilon}x$ is no longer small), the outer solution is

$$u_{\text{outer}} = g = \frac{1}{2 - x^2} \tag{11.22}$$

We now need to match the solutions at some intermediate value of x, and then obtain a uniformly valid approximate solution. In the limit as x approaches zero, g, the outer solution, approaches $1/2$, while when $x = 1$, the boundary condition at $x = 1$, $u(1) = 1$, is satisfied. If we choose $A = 0$ and $B = -1/2$, the limit as $\varepsilon \to 0$ for the inner solution and $\varepsilon \to 0$ for the outer solution yields

$$\lim_{\varepsilon \to 0} \left(\frac{1}{2} - \frac{1}{2} e^{-\sqrt{2/\varepsilon}x} \right) = \lim_{x \to 0} \frac{1}{2 - x^2} = \frac{1}{2} \tag{11.23}$$

The uniformly valid solution is given by the sum of the inner and outer solutions minus the common part.

$$u_0 = f_0 + g_0 - \frac{1}{2} = \frac{1}{2 - x^2} - \frac{1}{2} e^{-x\sqrt{2/\varepsilon}} \tag{11.24}$$

Example 11.3 (a problem from Carrier):   Consider now the differential equation

$$\varepsilon \frac{d^2u}{dx^2} - (2 - x^2)u = -1 \tag{11.25}$$

$$u(-1) = u(1) = 0$$

As in the above example, if ε were identically zero, the solution would be

$$u = \frac{1}{2 - x^2} \tag{11.26}$$

However, this solution cannot satisfy either of the two boundary conditions and, therefore, cannot be an accurate solution near either of the two boundaries. We conclude that there are boundary layers near the boundaries where the solution changes rapidly in order to satisfy the boundary conditions. Taking the above function to be an outer solution that is valid away from the boundaries, we assume that an approximate solution takes the form

$$u_0 + w(x, \varepsilon) + v(x, \varepsilon) \tag{11.27}$$

where $w(x, \varepsilon)$ vanishes rapidly away from $x = -1$ and $v(x, \varepsilon)$ vanishes rapidly away from $x = +1$. Near $x = -1$ it makes sense to define a new variable

$$\xi = (1 + x)\vartheta(\varepsilon) \tag{11.28}$$

and

$$w = W(\xi) \tag{11.29}$$

(The choice shifts the origin of the variable such that $x = -1$ corresponds to $\xi = 0$.)
Similarly, we define

$$\eta = (1 - x)\psi(\varepsilon) \tag{11.30}$$

and

$$v = V(\eta) \tag{11.31}$$

so that

$$u = u_0(x) + W(\xi) + V(\eta) \tag{11.32}$$
$$\varepsilon u_{0_{xx}} + \varepsilon \vartheta^2(\varepsilon) W_{\xi\xi} + \varepsilon \psi^2(\varepsilon) V_{\eta\eta} - (2 - x^2)[u_0 + W(\xi) + V(\eta)] = -1 \tag{11.33}$$

Near $\xi = 0$, and $\eta = 0$, $\varepsilon u_{0_{xx}} = O(\varepsilon)$ and if we choose

$$\vartheta(\varepsilon) = \psi(\varepsilon) = \sqrt{\varepsilon} \tag{11.34}$$

We find that near $\xi = 0$ or $\eta = 0$

$$W_{\xi\xi} - (2 - x^2)W(\xi) + V_{\eta\eta} - (2 - x^2)V(\eta) = 0 \tag{11.35}$$

to order ε.

Near $x = -1$ or $x = +1(2 - x^2)$ is of order 1.
Near $\xi = 0$, we consider the equation

$$W_{\xi\xi} - W = 0 \tag{11.36}$$

with solution satisfying the boundary condition $W(0) = -1$, with a bounded solution. The result is

$$W = -e^{-(1+x)/\sqrt{\varepsilon}} \tag{11.37}$$

By symmetry,

$$V = -e^{-(1-x)/\sqrt{\varepsilon}} \tag{11.38}$$

The complete solution is then

$$u = \frac{1}{2 - x^2} - \exp\left(-\frac{x+1}{\sqrt{\varepsilon}}\right) - \exp\left(-\frac{1-x}{\sqrt{\varepsilon}}\right) \tag{11.39}$$

Example 11.4   Consider now a slightly more general linear second order differential equation

$$\varepsilon y'' + a(x)y' + b(x)y = 0, \quad 0 < x < 1, \quad y(0) = 0, \quad y(1) = 1 \qquad (11.40)$$

where $\varepsilon << 1$ and both $a(x)$ and $b(x)$ are continuously differentiable.

It should be clear to the reader that a regular perturbation approach would lead to elimination of the second derivative term, and would not lead to a useful result. Example 11.1 hints that we should choose a new variable as

$$\xi = \varepsilon^{-1}g(x), \quad g(0) = 0 \qquad (11.41)$$

Thus,

$$y = y(x, \xi, \varepsilon)$$
$$y' = y_x + y_\xi \varepsilon^{-1}g'^2 \qquad (11.42)$$
$$y'' = y_{xx} + 2y_{x\xi}\varepsilon^{-1}g' + y_\xi \varepsilon^{-1}g'' + y_{\xi\xi}\varepsilon^{-2}g'^2$$

Here the subscripts represent differentiation with respect to the subscripted variable.

Substituting into (11.40),

$$g'^2 y_{\xi\xi} + ag'y_\xi + \varepsilon(2y_{x\xi}g' + y_\xi g'' + ay_x + +by) + \varepsilon^2 y_{xx} = 0$$
$$y(0, 0, \varepsilon) = 0, \, y\left(1, \varepsilon^{-1}g(1), \varepsilon\right) = 1 \qquad (11.43)$$

Now we assume that

$$y(x, \xi, \varepsilon) = Y_0(x, \xi) + \varepsilon Y_1(x, \xi) + \dots \qquad (11.44)$$

Equating to zero terms of successive powers of ε we find

$$g'^2 Y_{0,\xi\xi} + ag'Y_{0,\xi} = 0, \quad Y_0(0, 0) = 0, \quad Y_0(1, \varepsilon^{-1}g(1)) = 1 \qquad (11.45)$$

$$g'^2 Y_{1,\xi\xi} + ag'Y_{1,\xi} + 2g'Y_{0,x\xi} + g''Y_{0,\xi} + aY_{0,x} + bY_0 = 0,$$
$$Y_1(0, 0) = 0, \quad Y_1(1, \varepsilon^{-1}g(1)) = 0 \qquad (11.46)$$

The zeroth order solutions simplifies if we let

$$a(x) = g', \quad g(x) = \int_0^x a(x)\, dx \qquad (11.47)$$

Solving (11.45),

$$Y_0(x, \xi) = A_0(x) + B_0(x)\exp(-\xi) \qquad (11.48)$$

The first boundary condition in (11.45) gives

$$A_0(0) + B_0(0) = 0 \qquad (11.49)$$

At $x = 1$, $\exp(-\varepsilon^{-1} \int_0^1 a(x)dx)$, so that $\exp(-\xi)$ is transcendentally small. Thus,

$$A_0(1) = 1 \tag{11.49a}$$

The differential equation for Y_1 can now be written

$$Y_{1,\xi\xi} + Y_{1,\xi} = \exp^{-\xi} \left[\frac{B_0'}{a} + \frac{(a' - b) B_0}{a^2} \right] - \left(\frac{A_0'}{a} + \frac{bA_0}{a^2} \right) \tag{11.50}$$

If the right-hand side of this equation is nonzero, the solution will involve terms like $F(x)\xi e - \xi + G(x)\xi$. These are called resonant terms, and since $\xi = g(x)/\varepsilon$ and we are interested in the case where $\varepsilon \to 0$ the right-hand side of (11.50) must be identically zero.

Thus,

$$a A_0' + b A_0 = 0$$

and $$\tag{11.51}$$

$$a B_0' + (a' - b) B_0 = 0$$

The first of these has the solution

$$A_0(x) = \exp \int_x^1 [b(t)a(t)] \, dt \tag{11.52}$$

while the second yields

$$B_0(x) = [C_0/a(x)] \exp \left[\int_0^x (b(t)/a(t)) \, dt \right] \tag{11.53}$$

C_0 can be determined by applying (11.49) with A_0 given by (11.52) and we find that

$$C_0 = - a(0) \exp \int_0^1 [b(t)/a(t)] \, dt \tag{11.54}$$

The zeroth order solution is

$$Y_0(x, \xi) = \exp \int_x^1 [b(t)/a(t)] \, dt$$
$$- \frac{a(0)}{a(x)} e^{-\xi} \exp \int_0^1 [b(t)/a(t)] \, dt \exp \left[\int_0^x [b(t)/a(t)] \, dt \right] \tag{11.55}$$

Example 11.5   In some cases boundary layers occur in the interior of the solution domain. Such is the case with the following equation. Consider

$$\varepsilon^2 y'' + 2xy' - 2y = 0, \quad -1 < x < 1, \quad y(-1) = A, \quad y(1) = B \tag{11.56}$$

With $\varepsilon = 0$ identically, the reduced equation yields $y = Cx$, and the boundary condition at $x = -1$ requires that $C = -A$. On the other hand, applying the boundary condition at $x = 1$ gives $C = B$. The implication is that

$$y = -Ax, \quad x < 0$$
$$y = Bx, \quad x > 0$$

(11.57)

From this we deduce that at $x = 0$ there is a cusp, a point where the first derivative is discontinuous. We look for a solution near $x = 0$ in which the leading term is not negligible. Let

$$\xi = x/\varepsilon$$

(11.58)

Then

$$\frac{d^2 y}{d\xi^2} + 2\xi \frac{dy}{d\xi} - 2y = 0, \quad -\varepsilon^{-1} < \xi < \varepsilon^{-1}$$

(11.59)

One solution of this equation is clearly $y = \xi$. The other is then of the form (using variation of parameters)

$$y = \xi u(\xi)$$

(11.60)

Substituting into (11.59)

$$\xi u'' + 2(1 + \xi^2)u' = 0$$

(11.61)

and integrating once,

$$u' = c_1 \xi^{-2} \exp\left(-\xi^2\right)$$

(11.62)

A second integration gives

$$u = c_2 + c_1 \int_\xi^\infty \frac{\exp(-t^2)}{t^2} \, dt$$

(11.63)

Integration by parts gives

$$u = C + D\left[\frac{1}{\sqrt{\pi}} \exp\left(-\xi^2\right) + \frac{2}{\sqrt{\pi}} \int_\xi^\infty \exp\left(-t^2\right) dt\right]$$
$$= C + D\left[\frac{1}{\sqrt{\pi}} \exp\left(-\xi^2\right) + \mathrm{erf}(\xi)\right]$$

(11.64)

where $\mathrm{erf}(\xi)$ is the error function.

Now

$$y = Ex + F\left[\frac{\varepsilon}{\sqrt{\pi}} \exp\left(-x^2/\varepsilon^2\right) + x \, \mathrm{erf}(x/\varepsilon)\right]$$

(11.65)

Imposing the boundary conditions given in (11.56) and neglecting terms multiplied by ε,

$$y = x\left[\frac{B - A}{2} + \frac{B + A}{2} \mathrm{erf}(x/\varepsilon)\right] - \frac{B + A}{2\sqrt{\pi}} \varepsilon \exp\left(-x^2/\varepsilon^2\right)$$

(11.66)

Example 11.6 Finally, we consider a partial differential equation that describes the motion of the fluid in semicircular can of the human ear. The symbol u represents the velocity of the fluid inside the semicircular can relative to the wall of the canal. For the case of an impulse response (a sudden movement of the head) the equation describing the velocity of the fluid can be written

$$\frac{\partial u}{\partial t} + \left(1 + \frac{\gamma}{\beta}\right) \delta(t) = \frac{1}{r} \frac{\partial}{\partial r}\left(r\left(\frac{\partial u}{\partial r}\right)\right) - \varepsilon \int_0^t \int_0^1 u\, r\, dr\, dt \qquad (11.67)$$

The symbols γ and β represent properties of the canal whereas ε also contains properties of the canal fluid. If you are interested in the derivation of (11.67) you can find it in the article by van Buskirk, Watts and Liu found at the end of this chapter. For the human ear γ/β is of order unity and ε is very small, of order 0.02.

The boundary and initial conditions state that the velocity is zero at the wall ($r = 1$), the slope of the velocity is zero at the midpoint of the channel ($r = 0$) and the velocity is initially zero ($t = 0$).

$$u(1, t) = u(r, 0) = \frac{\partial u}{\partial r}(0, t) = 0 \qquad (11.68)$$

For *very small values of time* the regular expansion

$$u = u_0 + \varepsilon u_1 + \varepsilon^2 u_2 + \dots. \qquad (11.69)$$

leads to the following partial differential equation for u_0.

$$\frac{\partial u_o}{\partial t} + \left(1 + \frac{\gamma}{\beta}\right) \delta(t) = \frac{1}{r} \frac{\partial}{\partial r}\left(r \frac{\partial u_0}{\partial r}\right) \qquad (11.70)$$

The solution is easily obtained by Laplace transforms.

$$u_0 = -2 \left(1 + \frac{\gamma}{\beta}\right) \sum_{n=1}^{\infty} \frac{\exp\left(-\lambda_n^2 t\right) J_0\left(\lambda_n r\right)}{\lambda_n J_1\left(\lambda_n\right)} \qquad (11.71)$$

It is useful to determine the volume flow rate integral as

$$V = \int_0^t \int_0^1 u\, r\, dr\, dt \qquad (11.72)$$

This quantity represents the total displacement of the fluid. In the present case the integration is easily performed with the result:

$$V_0 = -2 \left(1 + \frac{\gamma}{\beta}\right) \sum_{n=1}^{\infty} \frac{1 - \exp\left(-\lambda_n^2 t\right)}{\lambda_n^4} \qquad (11.73)$$

Note now why the present system is singular. It is simply because for small values of time (and ε) the integral term in (11.67) can be neglected as a zeroth approximation, while for larger times, regardless of the (non-zero) value of ε, the integral term can be dominant.

For *large times* the velocity will be changing very slowly. We then let $\tau = \varepsilon t$, the stretched independent variable. Substituting into (11.67) we find

$$\varepsilon \frac{\partial u}{\partial \tau} + \left(1 + \frac{\gamma}{\beta}\right) \delta\left(\tau/\varepsilon\right) = \frac{1}{r}\left(r\frac{\partial u}{\partial r}\right) - \int_0^\tau \int_0^1 u r \, dr \, d\tau \tag{11.74}$$

The integral can now be written

$$\int_0^\nu \int_0^1 u r \, dr \, d\tau + \int_\nu^\tau \int_0^1 u r \, dr \, d\tau \tag{11.75}$$

When $\varepsilon << \nu(\varepsilon) << 1$, and if ε is sufficiently small then for all practical purposes the first of these integrals can be written

$$\int_0^\nu \int_0^1 u r \, dr \, d\tau = -2\left(1 + \frac{\gamma}{\beta}\right)\varepsilon \sum_{n=1}^\infty \frac{1}{\lambda_n^4} \tag{11.76}$$

so that to a zeroth approximation (11.74) can be written

$$\frac{1}{r}\frac{\partial}{\partial r}\left(r\frac{\partial u_0}{\partial r}\right) = \int_{\nu(\varepsilon)}^\tau \int_0^1 u_0 r \, dr \, d\tau - 2\left(1 + \frac{\gamma}{\beta}\right)\varepsilon \sum_{n=1}^\infty \left(\frac{1}{\lambda_n^4}\right) \tag{11.77}$$

Equation (11.77) is easily solved using Laplace transforms and the result is

$$u_0 = \frac{\varepsilon}{4}\sum_{n=1}^\infty \left(\frac{2\left(1 + \gamma/\beta\right)}{\lambda_n^4}\right)\left(1 - r^2\right)\exp\left(-\varepsilon t / 16\right) \tag{11.78}$$

The corresponding volume displacement is

$$V_0 = \int_{\nu(\varepsilon)}^t \int_0^1 u_0 r dr \, dt = \left[\exp(-\nu(\varepsilon)) - \exp(t\varepsilon/16)\right]\sum_{n=1}^\infty \frac{2\left(1 + \gamma/\beta\right)}{\lambda_n^4} \tag{11.79}$$

and since we have assumed that $\nu(\varepsilon) << 1$

$$V_0 = \left[1 - \exp(t\varepsilon/16)\right]\sum_{n=1}^\infty \frac{2(1 + \gamma/\beta)}{\lambda_n^4} \tag{11.80}$$

This is the solution for large values of time. Equations (11.73) and (11.80) can now be combined to give the uniformly valid zeroth approximation.

$$V_0 = \sum_{n=1}^\infty \frac{2\left(1 + \gamma/\beta\right)}{\lambda_n^2}\left[\exp\left(-\lambda_n^2 t\right) - \exp(-t\varepsilon/16)\right] \tag{11.81}$$

Problems

1. Find an approximate solution for small ε.

   (a) $\varepsilon y' + y = 0,\quad y(0) = 1$

   (b) $\varepsilon y'' + (1 + \varepsilon)y' + y = 0,\quad y(0) = 0,\quad y(1) = 1$

   (c) Show that the exact solution of (11.25) is

   (d) Consider

$$\varepsilon^2 y'' + y' + y + x = 1,\quad 0 < x < 1,\quad y(0) = 0,\quad y(1) = 0.$$

Noting that when ε is identically zero the solution is $y = 2 - x$, neither boundary condition can be satisfied. Singularities (boundary layers) occur at both $x = 0$ and $x = 1$. Find the zeroth order approximate solution.

   (e) Graph Equations (11.39) and (11.66)

REFERENCES

[1] Carrier, G. F. and C. E, Pearson, *Ordinary Differential Equations*, Blaisdell Pub. Co., Waltham, Mass. Cited on page(s)

[2] Van Buskirk, W. C., R. G. Watts, and Y. K. Liu, "The Fluid Mechanics of the Semicircular Canals," *J. Fluid Mech.*, vol. 78, Part 1, pp. 87–98, 1976. Cited on page(s)

[3] Simmonds, J. G. and J. E. Mann, *A First Look at Perturbation Theory*, Robert E. Krieger Pub. Co., Melebar, Florida, 1986. Cited on page(s)

[4] Ali Hasan Nayfeh, *Perturbation Methods*, Wiley, Weinham, Germany, 2007, 420 pp. Cited on page(s)

[5] Alan W. Bush, *Perturbation Methods for Engineers and Scientists*, CRC Press, 1992, 303 pp. Cited on page(s)

Appendix A: The Roots of Certain Transcendental Equations

TABLE A.1: The first six roots, † α_n, of

$$\alpha \tan \alpha + C = 0.$$

C	α_1	α_2	α_3	α_4	α_5	α_6
0	0	3.1416	6.2832	9.4248	12.5664	15.7080
0.001	0.0316	3.1419	6.2833	9.4249	12.5665	15.7080
0.002	0.0447	3.1422	6.2835	9.4250	12.5665	15.7081
0.004	0.0632	3.1429	6.2838	9.4252	12.5667	15.7082
0.006	0.0774	3.1435	6.2841	9.4254	12.5668	15.7083
0.008	0.0893	3.1441	6.2845	9.4256	12.5670	15.7085
0.01	0.0998	3.1448	6.2848	9.4258	12.5672	15.7086
0.02	0.1410	3.1479	6.2864	9.4269	12.5680	15.7092
0.04	0.1987	3.1543	6.2895	9.4290	12.5696	15.7105
0.06	0.2425	3.1606	6.2927	9.4311	12.5711	15.7118
0.08	0.2791	3.1668	6.2959	9.4333	12.5727	15.7131
0.1	0.3111	3.1731	6.2991	9.4354	12.5743	15.7143
0.2	0.4328	3.2039	6.3148	9.4459	12.5823	15.7207
0.3	0.5218	3.2341	6.3305	9.4565	12.5902	15.7270
0.4	0.5932	3.2636	6.3461	9.4670	12.5981	15.7334
0.5	0.6533	3.2923	6.3616	9.4775	12.6060	15.7397
0.6	0.7051	3.3204	6.3770	9.4879	12.6139	15.7460
0.7	0.7506	3.3477	6.3923	9.4983	12.6218	15.7524
0.8	0.7910	3.3744	6.4074	9'5087	12.6296	15.7587

TABLE A.1: (*continue*)

			$\alpha \tan \alpha + C = 0.$			
C	α_1	α_2	α_3	α_4	α_5	α_6
0.9	0.8274	3.4003	6.4224	9.5190	12.6375	15.7650
1.0	0.8603	3.4256	6.4373	9.5293	12.6453	15.7713
1.5	0.9882	3.5422	6.5097	9.5801	12.6841	15.8026
2.0	1.0769	3.6436	6.5783	9.6296	12.7223	15.8336
3.0	1.1925	3.8088	6.7040	9.7240	12.7966	15.8945
4.0	1.2646	3.9352	6.8140	9.8119	12.8678	15.9536
5.0	1.3138	4.0336	6.9096	9.8928	12.9352	16.0107
6.0	1.3496	4.1116	6.9924	9.9667	12.9988	16.0654
7.0	1.3766	4.1746	7.0640	10.0339	13.0584	16.1177
8.0	1.3978	4.2264	7.1263	10.0949	13.1141	16.1675
9.0	1.4149	4.2694	7.1806	10.1502	13.1660	16.2147
10.0	1.4289	4.3058	7.2281	10.2003	13.2142	16.2594
15.0	1.4729	4.4255	7.3959	10.3898	13.4078	16.4474
20.0	1.4961	4.4915	7.4954	10.5117	13.5420	16.5864
30.0	1.5202	4.5615	7.6057	10.6543	13.7085	16.7691
40.0	1.5325	4.5979	7.6647	10.7334	13.8048	16.8794
50.0	1.5400	4.6202	7.7012	10.7832	13.8666	16.9519
60.0	1.5451	4.6353	7.7259	10.8172	13.9094	17.0026
80.0	1.5514	4.6543	7.7573	10.8606	13.9644	17.0686
100.0	1.5552	4.6658	7.7764	10.8871	13.9981	17.1093
∞	1.5708	4.7124	7.8540	10.9956	14.1372	17.2788

† The roots of this equation are all real if $C > 0$.

TABLE A.2: The first six roots, † α_n, of

$$\alpha \cot\alpha + C = 0.$$

C	α_1	α_2	α_3	C	α_1	α_2
−1.0	0	4.4934	7.7253	10.9041	14.0662	17.2208
−0.995	0.1224	4.4945	7.7259	10.9046	14.0666	17.2210
−0.99	0.1730	4.4956	7.7265	10.9050	14.0669	17.2213
−0.98	0.2445	4.4979	7.7278	10.9060	14.0676	17.2219
−0.97	0.2991	4.5001	7.7291	10.9069	14.0683	17.2225
−0.96	0.3450	4.5023	7.7304	10.9078	14.0690	17.2231
−0.95	0.3854	4.5045	7.7317	10.9087	14.0697	17.2237
−0.94	0.4217	4.5068	7.7330	10.9096	14.0705	17.2242
−0.93	0.4551	4.5090	7.7343	10.9105	14.0712	17.2248
−0.92	0.4860	4.5112	7.7356	10.9115	14.0719	17.2254
−0.91	0.5150	4.5134	7.7369	10.9124	14.0726	17.2260
−0.90	0.5423	4.5157	7.7382	10.9133	14.0733	17.2266
−0.85	0.6609	4.5268	7.7447	10.9179	14.0769	17.2295
−0.8	0.7593	4.5379	7.7511	10.9225	14.0804	17.2324
−0.7	0.9208	4.5601	7.7641	10.9316	14.0875	17.2382
−0.6	1.0528	4.5822	7.7770	10.9408	14.0946	17.2440
−0.5	J.1656	4.6042	7.7899	10.9499	14.1017	17.2498
−0.4	1.2644	4.6261	7.8028	10.9591	14.1088	17.2556
−0.3	1.3525	4.6479	7.8156	10.9682	14.1159	17.2614
−0.2	1.4320	4.6696	7.8284	10.9774	14.1230	17.2672
−0.1	1.5044	4.6911	7.8412	10.9865	14.1301	17.2730
0	1.5708	4.7124	7.8540	10.9956	14.1372	17.2788
0.1	1.6320	4.7335	7.8667	11.0047	14.1443	17.2845
0.2	1.6887	4.7544	7.8794	11.0137	14.1513	17.2903
0.3	1.7414	4.7751	7.8920	11.0228	14.1584	17.2961
0.4	1.7906	4.7956	7.9046	11.0318	14.1654	17.3019

TABLE A.2: *(continue)*

$$\alpha \cot\alpha + C = 0.$$

C	α_1	α_2	α_3	C	α_1	α_2
0.5	1.8366	4.8158	7.9171	11.0409	14.1724	17.3076
0.6	1.8798	4.8358	7.9295	11.0498	14.1795	17.3134
0.7	1.9203	4.8556	7.9419	11.0588	14.1865	17.3192
0.8	1.9586	4.8751	7.9542	11.0677	14.1935	17.3249
0.9	1.9947	4.8943	7.9665	11.0767	14.2005	17.3306
1.0	2.0288	4.9132	7.9787	11.0856	14.2075	17.3364
1.5	2.1746	5.0037	8.0385	1 J.1296	14.2421	17.3649
2.0	2.2889	5.0870	8.0962	1 J.1727	14.2764	17.3932
3.0	2.4557	5.2329	8.2045	11.2560	14.3434	17.4490
4.0	2.5704	5.3540	8.3029	11.3349	14.4080	17.5034
5.0	2.6537	5.4544	8.3914	11.4086	14.4699	17.5562
6.0	2.7165	5.5378	8.4703	11.4773	14.5288	17.6072
7.0	2.7654	5,6078	8.5406	11.5408	14.5847	17.6562
8.0	2.8044	5.6669	8.6031	11.5994	14.6374	17.7032
9.0	2.8363	5.7172	8.6587	11.6532	14.6870	17.7481
10.0	2.8628	5.7606	8.7083	11.7027	14.7335	17.7908
15.0	2.9476	5.9080	8.8898	11.8959	14.9251	17.9742
20.0	2.9930	5.9921	9.0019	12.0250	15.0625	18.1136
30.0	3.0406	6.0831	9.1294	12.1807	15.2380	18.3018
40.0	3.0651	6.1311	9.1987	12.2688	15.3417	18.4180
50.0	3.0801	6.1606	9.2420	12.3247	15.4090	18.4953
60.0	3.0901	6.1805	9.2715	12.3632	15.4559	18.5497
80.0	3.1028	6.2058	9.3089	12.4124	15.5164	18.6209
100.0	3.1105	6.2211	9.3317	12.4426	15.5537	18.6650
∞	3.1416	6.2832	9.4248	12.5664	15.7080	18.8496

† The roots of this equation are all real if $C > -1$. These negative values of C arise in connection with the sphere, §9.4.

TABLE A.3: The first six roots α_n, of

$$\alpha J_1(\alpha) - C J_0(\alpha) = 0$$

C	α_1	α_2	α_3	α_4	α_5	α_6
0	0	3.8317	7.0156	10.1735	13.3237	16.4706
0.01	0.1412	3.8343	7.0170	10.1745	13.3244	16.4712
0.02	0.1995	3.8369	7.0184	10.1754	13.3252	16.4718
0.04	0.2814	3.8421	7.0213	10.1774	13.3267	16.4731
0.06	0.3438	3.8473	7.0241	10.1794	13.3282	16.4743
0.08	0.3960	3.8525	7.0270	10.1813	13.3297	16.4755
0.1	0.4417	3.8577	7.0298	10.1833	13.3312	16.4767
0.15	0.5376	3.8706	7.0369	10.1882	13.3349	16.4797
0.2	0.6170	3.8835	7.0440	10.1931	13.3387	16.4828
0.3	0 7465	3.9091	7.0582	10.2029	13.3462	16.4888
0.4	0.8516	3.9344	7.0723	10.2127	13.3537	16.4949
0.5	0.9408	3.9594	7.0864	10.2225	13.3611	16.5010
0.6	1.0184	3.9841	7.1004	10.2322	13.3686	16.5070
0.7	1.0873	4.0085	7.1143	10.2419	13.3761	16.5131
0.8	1.1490	4.0325	7.1282	10.2516	13.3835	16.5191
0.9	1.2048	4.0562	7.1421	10.2613	13.3910	16.5251
1.0	1.2558	4.0795	7.1558	10.2710	13.3984	16.5312
1.5	1.4569	4.1902	7.2233	10.3188	13.4353	16.5612
2.0	1.5994	4.2910	7.2884	10.3658	13.4719	16.5910
3.0	1.7887	4.4634	7.4103	10.4566	13.5434	16.6499
4.0	1.9081	4.6018	7.5201	10.5423	13.6125	16.7073
5.0	1.9898	4.7131	7.6177	10.6223	13.6786	16.7630
6.0	2.0490	4.8033	7.7039	10.6964	13.7414	16.8168
7.0	2.0937	4.8772	7.7797	10.7646	13.8008	16.8684
8.0	2.1286	4.9384	7.8464	10.8271	13.8566	16.9179
9.0	2.1566	4.9897	7.9051	10.8842	13.9090	16.9650
10.0	2.1795	5.0332	7.9569	10.9363	13.9580	17.0099
15.0	2.2509	5.1773	8.1422	11.1367	14.1576	17.2008
20.0	2.2880	5.2568	8.2534	11.2677	14.2983	17.3442
30.0	2.3261	5.3410	8.3771	11.4221	14.4748	17.5348
40.0	2.3455	5.3846	8.4432	11.5081	14.5774	17.6508
50.0	2.3572	5.4112	8.4840	11.5621	14.6433	17.7272
60.0	2.3651	5.4291	8.5116	11.5990	14.6889	17.7807
80.0	2.3750	5.4516	8.5466	11.6461	14.7475	17.8502
100.0	2.3809	5.4652	8.5678	11.6747	14.7834	17.8931
∞	2.4048	5.5201	8.6537	11.7915	14.9309	18.0711

Appendix B

In this table $q = (p/a)^{1/2}$; a and x are positive real; α, β, γ are unrestricted; k is a finite integer; n is a finite integer or zero; v is a fractional number; $1 \cdot 2 \cdot 3 \cdots n = n!$; $1 \cdot 3 \cdot 5 \cdots (2n-1) = (2n-1)!!$ $n\Gamma(n) = \Gamma(n+1) = n!$; $\Gamma(1) = 0! = 1$; $\Gamma(v)\Gamma(1-v) = \pi/\sin v\pi$; $\Gamma(\tfrac{1}{2}) = \pi^{1/2}$

NO.	TRANSFORM	FUNCTION
1	$\dfrac{1}{p}$	1
2	$\dfrac{1}{p^2}$	t
3	$\dfrac{1}{p^k}$	$\dfrac{t^{k-1}}{(k-1)!}$
4	$\dfrac{1}{p^{1/2}}$	$\dfrac{1}{(\pi t)^{1/2}}$
5	$\dfrac{1}{p^{3/2}}$	$2\left(\dfrac{t}{\pi}\right)^{\frac{1}{2}}$
6	$\dfrac{1}{p^{k+1/2}}$	$\dfrac{2^k}{\pi^{1/2}(2k-1)!!}t^{k-1/2}$
7	$\dfrac{1}{p^v}$	$\dfrac{t^{v-1}}{\Gamma(v)}$
8	$p^{1/2}$	$-\dfrac{1}{2\pi^{1/2}t^{5/2}}$
9	$p^{3/2}$	$\dfrac{3}{4\pi^{1/2}t^{5/2}}$
10	$p^{k-1/2}$	$\dfrac{(-1)^k(2k-1)!!}{2^k\pi^{1/2}t^{k+1/2}}$
11	p^{n-v}	$\dfrac{t^{v-n-1}}{\Gamma(v-n)}$
12	$\dfrac{1}{p+\alpha}$	$e^{-\alpha t}$
13	$\dfrac{1}{(p+\alpha)(p+\beta)}$	$\dfrac{e^{-\beta t} - e^{-\alpha t}}{\alpha - \beta}$
14	$\dfrac{1}{(p+\alpha)^2}$	$te^{-\alpha t}$

15	$\dfrac{1}{(p+\alpha)(p+\beta)(p+\gamma)}$	$\dfrac{(\gamma-\beta)e^{-\alpha t}+(\alpha-\gamma)e^{-\beta t}+(\beta-\alpha)e^{-\gamma t}}{(\alpha-\beta)(\beta-\gamma)(\gamma-\alpha)}$
16	$\dfrac{1}{(p+\alpha)^2(p+\beta)}$	$\dfrac{e^{-\beta t}-e^{-\alpha t}[1-(\beta-\alpha)t]}{(\beta-\alpha)^2}$
17	$\dfrac{1}{(p+\alpha)^3}$	$\dfrac{1}{2}t^2 e^{-\alpha t}$
18	$\dfrac{1}{(p+\alpha)^k}$	$\dfrac{t^{k-1}e^{-\alpha t}}{(k-1)!}$
19	$\dfrac{p}{(p+\alpha)(p+\beta)}$	$\dfrac{\alpha e^{-\alpha t}-\beta e^{-\beta t}}{\alpha-\beta}$
20	$\dfrac{p}{(p+\alpha)^2}$	$(1-\alpha t)e^{-\alpha t}$
21	$\dfrac{p}{(p+\alpha)(p+\beta)(p+\gamma)}$	$\dfrac{\alpha(\beta-\gamma)e^{-\alpha t}+\beta(\gamma-\alpha)e^{-\beta t}+\gamma(\alpha-\beta)e^{-\gamma t}}{(\alpha-\beta)(\beta-\gamma)(\gamma-\alpha)}$
22	$\dfrac{p}{(p+\alpha)^2(p+\beta)}$	$\dfrac{[\beta-\alpha(\beta-\alpha)t]e^{-\alpha t}-\beta e^{-\beta t}}{(\beta-\alpha)^2}$
23	$\dfrac{p}{(p+\alpha)^3}$	$t\left(1-\dfrac{1}{2}\alpha t\right)e^{-\alpha t}$
24	$\dfrac{\alpha}{p^2+\alpha^2}$	$\sin\alpha t$
25	$\dfrac{p}{p^2+\alpha^2}$	$\cos\alpha t$
26	$\dfrac{\alpha}{p^2-\alpha^2}$	$\sinh\alpha t$
27	$\dfrac{p}{p^2-\alpha^2}$	$\cosh\alpha t$
28	e^{-qx}	$\dfrac{x}{2(\pi\alpha t^3)^{1/2}}e^{-x^2/4\alpha t}$
29	$\dfrac{e^{-qx}}{q}$	$\left(\dfrac{\alpha}{\pi t}\right)^{1/2}e^{-x^2/4\alpha t}$
30	$\dfrac{e^{-qx}}{p}$	$\operatorname{erfc}\left[\dfrac{x}{2(\alpha t)^{1/2}}\right]$
31	$\dfrac{e^{-qx}}{qp}$	$2\left(\dfrac{\alpha t}{\pi}\right)^{1/2}e^{-x^2/4\alpha t}-x\operatorname{erfc}\left[\dfrac{x}{2(\alpha t)^{1/2}}\right]$
32	$\dfrac{e^{-qx}}{p^2}$	$\left(t+\dfrac{x^2}{2\alpha}\right)\operatorname{erfc}\left[\dfrac{x}{2(\alpha t)^{1/2}}\right]-x\left(\dfrac{t}{\alpha\pi}\right)^{1/2}e^{-x^2/4\alpha t}$
33	$\dfrac{e^{-qx}}{p^{1+n/2}}$	$\dfrac{(\gamma-\beta)e^{-\alpha t}+(\alpha-\gamma)e^{-\beta t}+(\beta-\alpha)e^{-\gamma t}}{(\alpha-\beta)(\beta-\gamma)(\gamma-\alpha)}$

34	$\dfrac{e^{-qx}}{p^{3/4}}$	$\dfrac{e^{-\beta t} - e^{-\alpha t}[1 - (\beta - \alpha)t]}{(\beta - \alpha)^2}$
35	$\dfrac{e^{-qx}}{q + \beta}$	$\dfrac{1}{2}t^2 e^{-\alpha t}$
36	$\dfrac{e^{-qx}}{q(q + \beta)}$	$\dfrac{t^{k-1} e^{-\alpha t}}{(k - 1)!}$
37	$\dfrac{e^{-qx}}{p(q + \beta)}$	$\dfrac{\alpha e^{-\alpha t} - \beta e^{-\beta t}}{\alpha - \beta}$
38	$\dfrac{e^{-qx}}{qp(q + \beta)}$	$(1 - \alpha t)e^{-\alpha t}$
39	$\dfrac{e^{-qx}}{q^{n+1}(q + \beta)}$	$\dfrac{\alpha(\beta - \gamma)e^{-\alpha t} + \beta(\gamma - \alpha)e^{-\beta t} + \gamma(\alpha - \beta)e^{-\gamma t}}{(\alpha - \beta)(\beta - \gamma)(\gamma - \alpha)}$
40	$\dfrac{e^{-qx}}{(q + \beta)^2}$	$\dfrac{[\beta - \alpha(\beta - \alpha)t]e^{-\alpha t} - \beta e^{-\beta t}}{(\beta - \alpha)^2}$
41	$\dfrac{e^{-qx}}{p(q + \beta)^2}$	$t\left(1 - \dfrac{1}{2}\alpha t\right)e^{-\alpha t}$
42	$\dfrac{e^{-qx}}{p - \gamma}$	$\sin \alpha t$
43	$\dfrac{e^{-qx}}{q(p - \gamma)}$	$\dfrac{1}{2}e^{\gamma t}\left(\dfrac{\alpha}{\gamma}\right)^{1/2}\left\{\begin{array}{l} e^{-x(\gamma/\alpha)^{1/2}}\operatorname{erfc}\left[\dfrac{x}{2(\alpha t)^{1/2}} - (\gamma t)^{1/2}\right] \\ +e^{x(\gamma/\alpha)^{1/2}}\operatorname{erfc}\left[\dfrac{x}{2(\alpha t)^{1/2}} + (\gamma t)^{1/2}\right] \end{array}\right\}$
44	$\dfrac{e^{-qx}}{(p - \gamma)^2}$	$\dfrac{1}{2}e^{\gamma t}\left\{\begin{array}{l} \left[t - \dfrac{x}{2(\alpha t)^{1/2}}\right]e^{-x(\gamma/\alpha)^{1/2}}\operatorname{erfc}\left[\dfrac{x}{2(\alpha t)^{1/2}} - (\gamma t)^{1/2}\right] \\ +\left[t + \dfrac{x}{2(\alpha t)^{1/2}}\right]e^{x(\gamma/\alpha)^{1/2}}\operatorname{erfc}\left[\dfrac{x}{2(\alpha t)^{1/2}} + (\gamma t)^{1/2}\right] \end{array}\right\}$
45	$\dfrac{e^{-qx}}{(p - \gamma)(q + \beta)},$ $\gamma \neq \alpha\beta^2$	$\dfrac{1}{2}e^{\gamma t}\left\{\begin{array}{l} \dfrac{\alpha^{1/2}}{\alpha^{1/2}\beta + \gamma^{1/2}}e^{-x(\gamma/\alpha)^{1/2}}\operatorname{erfc}\left[\dfrac{x}{2(\alpha t)^{1/2}} - (\gamma t)^{1/2}\right] \\ +\dfrac{\alpha^{1/2}}{\alpha^{1/2}\beta - \gamma^{1/2}}e^{x(\gamma/\alpha)^{1/2}}\operatorname{erfc}\left[\dfrac{x}{2(\alpha t)^{1/2}} + (\gamma t)^{1/2}\right] \end{array}\right\}$ $-\dfrac{\alpha\beta}{\alpha\beta^2 - \gamma}e^{\beta x + \alpha\beta^2 t}\operatorname{erfc}\left[\dfrac{x}{2(\alpha t)^{1/2}} + \beta(\alpha t)^{1/2}\right]$
46	$e^{x/p} - 1$	$\left(\dfrac{x}{t}\right)^{1/2} I_1\left[2(xt)^{1/2}\right]$
47	$\dfrac{1}{p}e^{x/p}$	$I_0\left[2(xt)^{1/2}\right]$
48	$\dfrac{1}{p^y}e^{x/p}$	$\left(\dfrac{t}{x}\right)^{(v-1)/2} I_{v-1}\left[2(xt)^{1/2}\right]$

49	$K_0(qx)$	$\dfrac{1}{2t}e^{-x^2/4\alpha t}$
50	$\dfrac{1}{p^{1/2}}K_{2v}(qx)$	$\dfrac{1}{2(\pi t)^{1/2}}e^{-x^2 8\alpha t}K_v\left(\dfrac{x^2}{8\alpha t}\right)$
51	$p^{v/2-1}K_v(qx)$	$x^{-v}\alpha^{v/2}2^{v-1}\int_{x^2/4\alpha t}^{\infty}e^{-u}u^{v-1}du$
52	$p^{v/2}K_v(qx)$	$\dfrac{x^v}{\alpha^{v/2}(2t)^{v+1}}e^{-x^2/4\alpha t}$
53	$\left[p-(p^2-x^2)^{1/2}\right]^v$	$v\dfrac{x^v}{t}I_v(xt)$
54	$e^{x\left[(p+\alpha)^{1/2}-(p+\beta)^{1/2}\right]^z}-1$	$\dfrac{x(\alpha-\beta)e^{-(\alpha+\beta)t/2}I_1\left[\frac{1}{2}(\alpha-\beta)t^{1/2}(t+4x)^{1/2}\right]}{t^{1/2}(t+4x)^{1/2}}$
55	$\dfrac{e^{x\left[p-(p+\alpha)^{1/2}(p+\beta)^{1/2}\right]}}{(p+\alpha)^{1/2}(p+\beta)^{1/2}}$	$e^{-(\alpha+\beta)(t+x)/2}I_0\left[\frac{1}{2}(\alpha-\beta)t^{1/2}(t+2x)^{1/2}\right]$
56	$\dfrac{e^{x\left[(p+\alpha)^{1/2}-(p+\beta)^{1/2}\right]^2}}{(p+\alpha)^{1/2}(p+\beta)^{1/2}\left[(p+\alpha)^{1/2}+(p+\beta)^{1/2}\right]^{2v}}$	$\dfrac{t^{v/2}e^{-(\alpha+\beta)t/2}I_v\left[\frac{1}{2}(\alpha-\beta)t^{1/2}(t+4x)^{1/2}\right]}{(\alpha-\beta)^v(t+4x)^{v/2}}$

Author Biography

Dr. Robert G. Watts is the Cornelia and Arthur L. Jung Professor of Mechanical Engineering at Tulane University. He holds a BS (1959) in mechanical engineering from Tulane, an MS(1960) in nuclear engineering from the Massachusetts Institute of Technology and a PhD (1965) from Purdue University in mechanical engineering. He spent a year as a Postdoctoral associate studying atmospheric and ocean science at Harvard University. He has taught advanced applied mathematics and thermal science at Tulane for most of his 43 years of service to that university.

Dr. Watts is the author of *Keep Your Eye on the Ball: The Science and Folklore of Baseball* (W. H. Freeman) and the editor of *Engineering Response to Global Climate Change* (CRC Press) and *Innovative Energy Strategies for CO2 Stabilization* (Cambridge University Press) as well as many papers on global warming, paleoclimatology energy and the physic of sport. He is a Fellow of the American Society of Mechanical Engineers.

Printed in the United States
by Baker & Taylor Publisher Services